Raman and SERS Investigations of Pharmaceuticals

Monica Baia · Simion Astilean · Traian Iliescu

Raman and SERS Investigations of Pharmaceuticals

Dr. Monica Baia
Faculty of Physics
Molec. Spectroscopy Dept.
Babes-Bolyai University
M. Kogalniceanu 1
400084 Cluj-Napoca
Romania
monib@phys.ubbcluj.ro

Dr. Traian Iliescu
Faculty of Physics
Molec. Spectroscopy Dept.
Babes-Bolyai University
M. Kogalniceanu 1
400084 Cluj-Napoca
Romania
ilitra@phys.ubbcluj.ro

Dr. Simion Astilean
Faculty of Physics
Molec. Spectroscopy Dept.
Babes-Bolyai University
M. Kogalniceanu 1
400084 Cluj-Napoca
Romania
sastil@phys.ubbcluj.ro

ISBN 978-3-540-78282-7   e-ISBN 978-3-540-78283-4

DOI 10.1007/978-3-540-78283-4

Library of Congress Control Number: 2008920953

Coverdesign: WMXDesign GmbH, Heidelberg

Printed on acid-free paper

9 8 7 6 5 4 3 2 1

springer.com

Preface

In the last years it has become obvious for most researchers that interdisciplinary research is the key to success in the future of science. Therefore, only the collaborative work of chemists, biologists, physicists, and so on will lead to further developments.

The present book gives an example of such an interdisciplinary work. Thus, some new derivatives have been prepared by chemists, and were analyzed by physicists in order to better understand their physical-chemical properties for upcoming tests performed by pharmacists.

The book is addressed to students and researchers in the faculties of physics, chemistry, medicine and pharmacy as well as to employers in research departments of companies that are interested in designing and testing new and different types of pharmaceuticals. This book illustrates several of our recent spectroscopic investigations performed by using Raman, infrared, and surface-enhanced Raman spectroscopy (SERS) on different drugs such as tranquilizers and sedatives, anti-inflammatory drugs, vitamins, drugs with antibacterial properties, etc. The book ends by presenting some recent results about the preparation and investigation of the SERS efficiency of a few ordered and disordered metallic nanostructured substrates. Because of the possibility of tuning their plasmonic response, the analysis of the adsorption of such pharmaceutical relevant molecules on this kind of SERS-active substrates could provide further insights into their adsorption behavior.

Observations and suggestions will be gratefully received and may be addressed to the authors.

Cluj-Napoca, January 2008   Monica Baia, Simion Astilean, and Traian Iliescu

About the Authors

Monica Baia was born in Cluj-Napoca, Romania in 1974. She studied physics at the Babes-Bolyai University of Cluj-Napoca, in Romania, and received her PhD degree in 2003 on the vibrational characterization of coordination and biologically active compounds by means of IR absorption, Raman and surface-enhanced Raman spectroscopy (SERS) under the supervision of Prof. Dr. Dr. h.c. Wolfgang Kiefer from the University of Würzburg, in Germany. In 2003 she began her academic career as a lecturer at the Molecular Spectroscopy Department of the Babes-Bolyai University. In October, 2005 she worked for a nine month period with Prof. Dr. Jürgen Popp's group as a postdoctoral research fellow at the University of Jena, Germany. In 2007 she received the In Hoc Signo Vinces prize of the National University Research Council of Romania. Her current research interests include the elucidation of the structure and adsorption behavior of pharmaceutical and biological relevant compounds by using Raman spectroscopy and SERS, and the study of the optical and structural properties of metallic ordered and disordered nanostructures for their further use as SERS substrates in biomolecular and pollutants detection. She has about 50 refereed journal publications.

Simion Astilean was born in Cornesti (Cluj), Romania in 1956. He studied physics at the Babes-Bolyai University of Cluj-Napoca, in Romania, and completed his PhD on the spectroscopic characterization of intermolecular proton transfer in 1993 under the supervision of Prof. Dr. Peter Trommsdorff at the Joseph Fourier University, Grenoble, France. He worked for many years as a visiting scientist in the UK, Israel, and France. Astilean is currently the head of the Molecular Spectroscopy Department and the Nanoscience and Nanotechnology Center at the Babes-Bolyai University. His current research focuses on the rational design, fabrication, and assembly of noble-metal nanoparticles with the aim of enabling novel methods for detection and sensing biomolecules via surface-enhanced Raman spectroscopy (SERS), surface-enhanced IR absorption (SEIRA), local surface plasmon resonance (LSPR), and surface-enhanced fluorescence (SEF). He has authored and co-authored about 100 refereed journal publications, two books, and two book chapters.

Traian Iliescu was born in Slatina, Romania in 1941. He studied physics at the Babes-Bolyai University of Cluj-Napoca, in Romania. He received his PhD degree in 1975 from the Babes-Bolyai University. Iliescu began his academic career in 1963 as assistant professor at the Chemical Physics Department, in the Faculty of Pharmacy in Cluj-Napoca. From 1970 onwards he has been present in the Physics faculty in Cluj-Napoca as assistant professor, and then as full professor. His first research interest was the Spolski effect in matrix at low temperatures of different aromatic compounds and then the application of Raman spectroscopy in molecular dynamics by band shape analysis, and the elucidation of the structure of different glass systems and surface-enhanced Raman spectroscopy on the compounds of pharmaceutical and biological interest. In 1990 he obtained the C. Miculescu prize of the Romanian Academy for a selection of papers related to molecular dynamics. He has authored and co-authored about 100 refereed journal publications and four books.

Contents

Abbreviations

FT	Fourier transform
SERS	Surface-enhanced Raman spectroscopy
SERRS	Surface-enhanced resonance Raman spectroscopy
DFT	Density functional theory
HF	Hartree–Fock
RHF	Restricted Hartree–Fock
10-I-10H-P-5-O	10-Isopentyl-10H-phenothiazine-5-oxide
10-I-10H-P-5,5-D	10-Isopentyl-10H-phenothiazine-5,5-dioxide
PhT	Phenothiazine
DCFNa	Diclofenac sodium
NSAIDs	Nonsteroidal anti-inflammatory drugs
UV	Ultraviolet
UV-vis	Ultraviolet-visible
βCD	β-Cyclodextrin
FWHM	Full width at half maximum
KBP	Potassium benzylpenicillin
THA	Trihydrate amoxicillin
RIV	Rivanol
NAD	Nicotinamide adenine dinucleotide
NADP	Nicotinamide adenine dinucleotide phosphate
2FF	2-Formylfuran
5-(4FP)-2FF	5-(4-Fluor-phenyl)-2-formylfuran
5-(4Br-P)-2FF	5-(4-Brom-phenyl)-2-formylfuran
9-PA	9-Phenylacridine
p-ATP	*p*-Aminothiophenol
EF	Enhancement factor
LSPR	Localized surface plasmon resonance

1 Introduction

Infrared and Raman spectroscopy are two of the most widely used techniques in the physical and natural sciences today. In 1800 Sir William Herschel, while studying the heating effect produced by various portions of the solar spectrum, established that it contained some form of radiant energy which could not be seen (Herschel 1800). A few years later, in 1840, his son Sir John Herschel was able to demonstrate the existence of infrared absorption and transmission bands by noting variations in the rate of evaporation of alcohol from blackened paper upon which the solar spectrum was projected (Kruse et al. 1962). By utilizing detectors, the science of infrared moved steadily ahead and the idea that infrared radiation was quite similar to visible light began to be accepted. The utility of infrared spectroscopy as a tool for identification of molecules and functional groups was realized by chemists in the late 1920s. Modern infrared spectroscopy started in the 1940s and 1950s with tremendous improvements in instrumentation, which put the technique at the head of physical and chemical research (Kruse et al. 1962).

The Raman effect also allows the observation of vibrational spectra providing information which complements those obtained by infrared spectroscopy. This effect had been repeatedly predicted. Lommel (Lommel 1878) described certain anomalies of fluorescence, the color of which is dependent on the nature of the sample and the frequency of the exciting radiation. Smekal (Smekal 1923), Kramers and Heisenberg (Kramers and Heisenberg 1925), Schrödinger (Schrödinger 1926), and Dirac (Dirac 1927) predicted the Raman effect by applying quantum mechanics to molecules. Raman was looking for the optical analogue of the Compton effect, when his co-workers Krishnan and Venkateswaran observed "modified scattering" of sunlight, which Raman identified as the Kramers-Heisenberg effect. A short paper titled "A New Type of Secondary Radiation" by Raman and Krishnan was submitted to Nature on February 16th, 1928 (Raman and Krishnan 1928). As mentioned above, the basic theory of the Raman effect was developed before its discovery. However, at this time, numerical calculations of the intensity of Raman lines were impossible, because this requires information on all eigenstates of a scattering system. Placzek (Placzek 1934) introduced a "semiclassical" approach in the form of his polarizability theory. This provided a basis for many other theoretical and experimental studies. The most important stimulus to the development of the Raman spectroscopy has been the laser, invented by Maiman in 1960 (Maiman 1960). During a short period the mercury arcs were replaced by really monochromatic and powerful light sources. At the same time, the photographic plates were replaced by photomultipliers, and scanning grating spectrometers replaced the prism spectrographs. Also, the introduction of double and triple monochromators, an elaborate sample technique (Kiefer 1977), and later the introduction of diode arrays and charge-coupled devices (CCDs) contributed considerably to the development of Raman spectroscopy. However, until about 1950, Raman spectroscopy was applied more often than infrared spectroscopy. After 1950, when automatically recording infrared spectrometers were introduced to the market, infrared spectroscopy became widely used in routine analysis.

Due to its non-destructive character, Raman spectroscopy represents, together with infrared absorption spectroscopy, one of the most useful tools for obtaining information about the structure and properties of molecules from their vibrational transitions, despite the fact that the direct assignment of the infrared or Raman bands of relatively complex species is rather complicated (Nakamoto 1997, Nafie 2001). Theoretical simulations can certainly assist in obtaining a deeper understanding of the vibrational spectra of complicated molecules. Recently, it was shown that density functional theory (DFT) methods are a powerful computational alternative to the conventional quantum chemical methods, since they are much less computationally demanding and take account of the effects of electron correlation (Parr and Yang 1989, Seminario and Politzer 1995).

However, the application of conventional Raman spectroscopy is limited by the weak intensity of the Raman scattered light and the appearance of fluorescence. One way to overcome these disadvantages is the use of surface-enhanced Raman spectroscopy (SERS) (Moskovits 1985, Vo-Dinh 1988, Campion and Kambhampati 1988). The existence of SERS was recognized thirty years ago and since then has been demonstrated to be a powerful analytical tool for the sensitive and selective detection of molecules adsorbed on nanostructured (i. e., roughened), coinage metal surfaces. SERS was first discovered by Fleischmann and coworkers in 1974, when they observed the strong Raman scattering of the pyridine molecules adsorbed on electrochemically roughened silver electrodes (Fleischmann et al. 1974). The scientific groups of Jeanmaire and Van Duyne (Jeanmaire and Van Duyne 1977) and Albrecht and Creighton (Albrecht and Creighton 1977) confirmed this enhancement phenomenon (up to 10^6) and attributed the effect to complex surface enhancement processes. As in the case of other scientific fields, the development of SERS with regard to detailed understanding, extensive application, and widespread acceptance was not continuous; to a certain extent, SERS has experienced its up and down periods.

The first progress period of SERS started immediately after its discovery and lasted until the mid 1980s. Throughout this time interval the research on SERS was largely populated by representatives of the condensed matter physics and chemical physics communities. The research activity was mainly focused on getting a mechanistic understanding of the 10^6 fold intensity enhancement observed for normal Raman scattering. It was stated (Moskovits 1985, Moskovits 1982, Gao et al. 1990) that the enhancement of the Raman signal can be considered as the product of two main contributions: an electromagnetic enhancement mechanism and a chemical or charge-transfer enhancement mechanism. The contribution of the electromagnetic mechanism to the total enhancement is of the order of 10^4, while the chemical mechanism participation is in the range of 10^2. Surface selection rules were also presented in this time. In their simplest form, and assuming no specific symmetry selection rules, the most intense bands are predicted as those from vibrations, which induce a polarization of the adsorbate electron cloud perpendicular to the metal surface (Creighton 1988, Moskovits and Suh 1984). This information can be used qualitatively to find out details about the angle formed between the adsorbed molecule and the metal surface. One should also emphasize that surface-enhanced resonance Raman scattering (SERRS) with combined SERS and RRS enhancement factors in the 10^9–10^{10} range was already known (Jeanmaire and Van Duyne 1977, Sequaris and Koglin 1985).

In the next decade, SERS developed into quite a mature field. During this period, the attention of the representative researchers of the condensed matter physics was turned away from SERS to other subjects, and their place was taken by researchers interested mainly in applications of SERS to problems from electrochemistry, heterogeneous catalysis, polymer science, the biochemistry of surface immobilized proteins, and many others (Kneipp et al. 1999, Kneipp et al. 2002). However, this period was not so exciting as the earlier one and therefore can be considered as a down period.

In striking contrast, in the last ten years the interest in SERS was completely revived, mainly because of the remarkable discovery of a single molecule by SERS (Kneipp et al. 1997, Nie and Emory 1997) Moreover, the developments achieved in nanoscience and nanotechnology have determined a strong interest in SERS. Nowadays, it is difficult to find a paper on nanoscale optical properties which does not recommend SERS as the first example for applications. Today, there is an astonishing research interest concerning how to control, manipulate, and amplify light on the nanometer length scale using the properties of the collective electronic excitations in noble metal films or nanoparticles, known as surface plasmons. The interactions between adsorbed molecules and plasmonic nanostructures (Van Duyne 2004) may possibly have a considerable impact on many applications, such as localized surface plasmon resonance spectroscopy for chemical and biological sensing, sub-wavelength optical microscopy, and nanolithography, as well as SERS (Srituravanich et al. 2004, Anderson et al. 2005a, Anderson et al. 2005b, Haes and Van Duyne 2002, Riboh et al. 2003, Haes et al. 2004a, Haes et al. 2004b, Haes and Van Duyne 2004, Haes et al. 2005).

Although the theoretical understanding of the mechanism of surface enhancement is not definite and still evolving, the experimental data accumulated in the last years has demonstrated SERS to be a sufficiently sensitive spectroscopic method for surface science, analytical and environmental applications, biomedicine, biophysics, and biochemistry (Cotton et al. 1991, Baker and Moore 2005,

Haynes et al. 2005, Dieringer et al. 2006, Rosi and Mirkin 2005, Kneipp et al. 2002).

In this book, we present several of our recent results concerning SERS investigations on molecules of pharmaceutical interest (Iliescu et al. 1994, Iliescu et al. 1997, Iliescu et al. 2000, Iliescu et al. 2001, Iliescu et al. 2002a, Iliescu et al. 2002b, Iliescu et al. 2002c, Iliescu et al. 2003a, Iliescu et al. 2003b, Iliescu et al. 2003–2004, Iliescu et al. 2004a, Iliescu et al. 2004b, Iliescu et al. 2004c, Iliescu et al. 2006, Bolboaca et al. 2002, Bolboaca et al. 2003a, Bolboaca et al. 2004a, Baia et al. 2004, Baia and Baia 2005, Leopold et al. 2005) followed by a few studies related to the testing of the SERS efficiency of newly designed SERS substrates that would contribute to the improvement of the Raman enhancement of the adsorbed molecules, and to the enlargement of the investigation possibilities of the adsorption behavior of such molecules by SERS (Baia et al. 2006a, Toderas et al. 2004a, Toderas et al. 2004b, Baia et al. 2005a, Baia et al. 2006b, Toderas et al. 2006a, Toderas et al. 2006b, Astilean et al. 2006a, Toderas et al. 2007, Bolboaca et al. 2003c, Astilean et al. 2004a, Bolboaca et al. 2004b, Astilean et al. 2003, Astilean et al 2004b, Baia et al. 2005b, Astilean et al. 2005, Baia et al. 2006c, Baia et al. 2006d, Astilean et al. 2006b, Baia et al. 2006e).

Why SERS on pharmaceuticals? From pharmacological studies it is known that each drug is specific to a certain human organ on which it is adsorbed on some special centers. The adsorption of the molecules on a metal surface can be seen to mimic this adsorption process. In these investigations, the silver or gold surface can serve as an artificial biological interface (Dryhurst 1977). Moreover, for a complete understanding of the action of various drugs, such as the derivatives discussed in this book, it is very important to know if the structure of the adsorbed molecules is the same as that of the free species, and also to establish whether or not the molecule-substrate interaction may be dependent on the pH value of the environmental solution.

Besides this introduction section, the book consists of other eight chapters. Prior to presenting the experimental results, the fundamentals of infrared, Raman, and surface-enhanced Raman spectroscopy are highlighted. A few details about the experimental measurements are also specified. Since the assignment of the vibrational modes of most of the investigated species was performed with the help of theoretical simulations, a short presentation of the employed methods is also included in this chapter.

The SERS investigations illustrated in the next chapters are focused on different kind of drugs: tranquilizers and sedatives, anti-inflammatory drugs, vitamins, drugs with anti-bacterial properties, etc. Since there is an increased interest in designing highly effective and controllable SERS-active substrates, a few newly-developed substrates that could contribute to a deeper understanding and knowledge of the adsorption behavior of various types of molecules of pharmaceutical and medical interest are presented in the eighth chapter. The conclusions drawn from all these investigations are summarized in the last chapter.

References

Albrecht MG, Creighton JA (1977) Anomalously intense Raman spectra of pyridine at a silver electrode. J Am Chem Soc 99:5215–5217

Anderson N, Hartschuh A, Cronin S, Novotny L (2005a) Nanoscale vibrational analysis of single-walled carbon nanotubes. J Am Chem Soc 127:2533–2537

Anderson N, Hartschuh A, Cronin S, Novotny L (2005b) Near-field Raman microscopy. Materials Today 8:50–54

Astilean S, Bolboaca M, Maniu D, Iliescu T (2003) Ordered metallic nanostructures for surface-enhanced Raman spectroscopy. Book of Abstracts of the 3rd Conference New Research Trends in Material Science, Constanta, 266

Astilean S, Bolboaca M, Maniu D, Iliescu T (2004a) Ordered metallic nanostructures for surface-enhanced Raman spectroscopy. Rom Rep Phys 56:346–351

Astilean S, Baia M, Maniu D, Pinzaru S, Iliescu T (2004b) Fabrication of ordered noble-metal nanostructures via nanosphere lithography and their investigation as effective substrates for surface-enhanced Raman spectroscopy. Book of Abstracts of the International Bunsen Discussion Meeting "Raman and IR spectroscopy in biology and medicine", Jena, 84

Astilean S, Baia M, Baia L, Farcau C, Toderas F (2005) Noble-metal films deposited on polystyrene colloidal crystal as effective substrate for surface-enhanced Raman spectroscopy. Book of Abstracts of the Surface Plasmon Photonics 2 Confererence, Graz, 124

Astilean S, Baia M, Maniu D, Baia L, Toderas F, Bica E, Iosin M, Popescu O, Craciun C, Barbu-Tudoran L, Socaciu C, Popp J, Baldeck PL (2006a) Bio-plasmonics: Sensing biomolecular interactions with gold nanoparticles. Book of Abstracts of the 2nd International Conference Advanced Spectroscopies on Biomedical and Nanostructured Systems, Cluj-Napoca, 61

Astilean S, Baia M, Baia L, Farcau C, Maniu D (2006b) Tunable surface-enhanced Raman scattering (SERS) from noble metal films deposited on polystyrene colloidal crystal and nanoparticle arrays fabricated by nanosphere litography. Meeting Digest of the EOS Topical Meeting on Molecular Plasmonic Devices, Engelberg, 74–76

Baia M, Toderas F, Mihut A, Baia L, Astilean S (2005a) Gold colloidal particles on functionalized glass substrate for optical sensing and SERS. Book of Abstracts of the 11th European Conference on Applications of Surface and Interface Analysis (ECASIA) Vienna, 161

Baia M, Baia L, Astilean S (2005b) Gold nanostructured films deposited on polystyrene colloidal crystal templates for surface-enhanced Raman spectroscopy. Chem Phys Lett 404:3–8

Baia M, Toderas F, Baia L, Popp J, Astilean S (2006a) Probing the enhancement mechanisms of SERS with p-aminothiophenol molecules adsorbed on self-assembled gold colloidal nanoparticles. Chem Phys Lett 422:127–132

Baia M, Toderas F, Baia L, Popp J, Astilean S (2006b) Monitoring the interplay of SERS enhancement mechanisms contribution with p-aminothiophenol adsorbed on self-assembled gold nanoparticles. Book of Abstracts of the 2nd International Conference Advanced Spectroscopies on Biomedical and Nanostructured Systems, Cluj-Napoca, 122

Baia M, Baia L, Popp J, Astilean S (2006c) Surface-enhanced Raman scattering efficiency of truncated tetrahedral Ag nanoparticle arrays mediated by electromagnetic couplings. Appl Phys Lett 88:143121-1-143121-3

Baia M, Baia L, Popp J, Astilean S (2006d) Ordered metallic nanostructures obtained by nanosphere lithography as tunable SERS-active substrates. Book of Abstracts of the 2nd International Conference Advanced Spectroscopies on Biomedical and Nanostructured Systems, Cluj-Napoca, 123

Baia L, Baia M, Popp J, Astilean S (2006e) Gold films deposited over regular arrays of polystyrene nanospheres as highly effective SERS substrates from visible to NIR. J Phys Chem B 110:23982–23986

Baia M, Baia L (2005) Infrared absorption, Raman and SERS investigations of 2,1-benzisoxazole. Studia Physica L2:113–122

Baia M, Baia L, Kiefer W, Popp J (2004) Surface-enhanced Raman scattering and density functional theoretical study of anthranil adsorbed on colloidal silver particles. J Phys Chem B 108:17491–17496

Baker GA, Moore DS (2005) Progress in plasmonic engineering of surface-enhanced Raman-scattering substrates toward ultra-trace analysis. Anal Bioanal Chem 382:1751–1770

Bolboaca M, Iliescu T, Kiefer W (2002) Fourier transform Raman and surface-enhanced Raman spectroscopy of some quinoline derivatives. J Raman Spectrosc 33:207–212

Bolboaca M, Iliescu T, Kiefer W (2004) Infrared absorption, Raman, and SERS investigations in conjunction with theoretical simulations on a phenothiazine derivative. Chem Phys 298:87–95

Bolboaca M, Iliescu T, Paizs Cs, Irimie FD, Kiefer W (2003a) Raman, infrared, and surface-enhanced Raman spectroscopy in combination with ab initio and density functional theory calculations on 10-Isopropyl-10H-phenothiazine-5-oxide. J Phys Chem A 107:1811–1818

Bolboaca M, Baia L, Chicinas I, Iliescu T, Astilean S (2003b) Optical and structural investigations of ordered metallic nanostructures for SERS experiments. Studia UBB Physica, XLVIII:357–359

Bolboaca M, Baia L, Chicinas I, Iliescu T, Astilean S (2003c) Optical and structural investigations of ordered metallic nanostructures for SERS experiments. Book of Abstracts of the 3rd Conference of Isotopic and Molecular Processes, Cluj-Napoca, 99

Campion A, Kambhampati P (1998) Surface enhanced Raman scattering. Chem Soc Rev 27:241–250

Cotton TM, Kim JH, Chumanov GD (1991) Application of surface-enhanced Raman spectroscopy to biological systems. J Raman Spectrosc 22:729–742

Creighton JA (1988) The selection rules for surface-enhanced Raman. In: Clark RH and Hester R (eds) Spectroscopy of Surface, Wiley & Sons, New York

Dieringer JA, McFarland AD, Shah NC, Stuart DA, Whitney AV, Yonzon CR. Young MA, Zhang X, Van Duyne RP (2006) Surface enhanced Raman spectroscopy: new materials, concepts, characterization tools, and applications. Faraday Discuss 132:9–26

Dirac PAM (1927) The quantum theory of dispersion. Proc Roy Soc A 114:710–728

Dryhurst CG (1977) Electrochemistry of biological molecules. Academic Press, New York

Fleischmann M, Hendra PJ, McQuillann AJ (1974) Raman spectra of pyridine adsorbed at a silver electrode. Chem Phys Lett 26:163–166

Gao X, Davies JP, Weaver MJ (1990) A test of surface selection rules for surface-enhanced Raman scattering: the orientation of adsorbed benzene and monosubstituted benzenes on gold. J Phys Chem 94:6858–6865

Haes AJ, Van Duyne RP (2002) A nanoscale optical biosensor: Sensitivity and selectivity of an approach based on the localized surface plasmon resonance spectroscopy of triangular silver nanoparticles. J Am Chem Soc 124:10596–10604

Haes AJ, Zou S, Schatz GC, Van Duyne RP (2004a) A nanoscale optical biosensor: The long range distance dependence of the localized surface plasmon resonance of noble metal nanoparticles. J Phys Chem B 108:109–116

Haes AJ, Zou S, Schatz GC, Van Duyne RP (2004b) Nanoscale optical biosensor: Short range distance dependence of the localized surface plasmon resonance of noble metal nanoparticles. J Phys Chem B 108:6961–6968

Haes AJ, Van Duyne RP (2004) A unified view of propagating and localized surface plasmon resonance biosensors. Anal Bioanal Chem 379:920–930

Haes AJ, Chang L, Klein WL, Van Duyne RP (2005) Detection of a biomarker for Alzheimer's disease from synthetic and clinical samples using a nanoscale optical biosensor. J Am Chem Soc 127:2264–2271

Haynes CL, McFarland AD, Van Duyne RP (2005) Surface enhanced Raman spectroscopy. Anal Chem 340:339–346

Herschel W (1800) Experiments on the refrangibility of the invisible rays of the Sun. Phil Trans Roy Soc 90:284–292

Iliescu T, Marian I, Misca R, Smarandache V (1994) Surface-enhanced Raman spectroscopy of 9-phenylacridine on silver sol. Analyst 119:567–570

Iliescu T, Cinta S, Astilean S, Bratu I (1997) pH influence on the Raman spectra of PP vitamin in silver sol. J Molec Struct 410–411:93–196

Iliescu T, Cinta S, Kiefer W (2000) FT-Raman and SERS spectra of rivanol in silver sol. Talanta 53:121–124

Iliescu T, Irimie FD, Bolboaca M, Paisz Cs, Kiefer W (2001) Raman spectra of 5 substituted-furan-2-carbaldehyde adsorbed on silver sol. Book of Abstracts of the 1st International Conference on Advanced Vibrational Spectroscopy (ICAVS), Turku, P171

Iliescu T, Irimie FD, Bolboaca M, Paisz Cs, Kiefer W (2002a) Vibrational spectroscopic investigations of 5-(4-fluor-phenyl)-furan-2 carbaldehyde. Vib Spectrosc 29:235–239

Iliescu T, Irimie FD, Bolboaca M, Paisz Cs, Kiefer W (2002b) Surface enhanced Raman spectroscopy of 5-(4-fluor-phenyl)-furan-2 carbaldehyde adsorbed on silver colloid. Vib Spectrosc 29:251–255

Iliescu T, Bolboaca M, Cinta-Pinzaru S, Pacurariu R, Maniu D, Ristoiu M, Kiefer W (2002c) Raman, surface-enhanced Raman spectroscopy and density functional theory studies of 2-formylfuran. Proceedings of the XVIIIth International Conference on Raman Spectroscopy (ICORS), Budapest, 311–312

Iliescu T, Bolboaca M, Astilean S, Maniu D, Kiefer W (2003a) Surface-enhanced Raman spectroscopy and theoretical investigations of diclofenac sodium. Book of Abstracts of the Third Conference of Isotopic and Molecular Processes, Cluj-Napoca, 47

Iliescu T, Bolboaca M, Pacurariu R, Maniu D, Kiefer W (2003b) Raman spectroscopy, surface-enhanced Raman spectroscopy and density functional theory studies of 2-formylfuran J. Raman Spectrosc 34:705–710

Iliescu T, Irimie FD, Baia M, Paizs Cs, Bratu I, Kiefer W (2003–2004) IR absorption, FT-Raman and SERS investigations together with DFT calculations on a furan-2-carbaldehyde derivative. Asian Chem Lett 7:213–220

Iliescu T, Baia M, Kiefer W (2004a) FT-Raman, surface-enhanced Raman spectroscopy and theoretical investigations of diclofenac sodium. Chem. Phys 298:167–174 a

Iliescu T, Baia M, Miclăuş V (2004b) A Raman spectroscopic study of the diclofenac sodium-β-cyclodextrin interaction. Eur J Pharma Sci 22:487–495

Iliescu T, Baia M, Miclăuş V, Kiefer W (2004c) A Raman spectroscopic study of the diclofenac sodium-β-cyclodextrin interaction. Proceedings of the XIXth International Conference on Raman Spectroscopy (ICORS), CSIRO Publishing, Gold Coast, Queensland, 470–471

Iliescu T, Baia M, Pavel I (2006) Raman and SERS investigations of potassium benzylpenicillin. J Raman Spectrosc 37:318–325

Jeanmaire DL, Van Duyne RP (1977) Surface Raman spectroelectrochemistry. Part I. Heterocyclic, aromatic, and aliphatic amines adsorbed on the anodized silver electrode. J. Electroanal Chem 84:1–20

Kiefer W (1977) Recent techniques in Raman spectroscopy. In: Clark RJH, Hester RE (eds) Advances in infrared and Raman spectroscopy, Vol. 3, London

Kneipp K, Wang Y, Kneipp H, Perelman LT, Itzkan I, Dasari RR, Feld MS (1997) Single molecule detection using surface-enhanced Raman scattering Phys Rev Lett 78:1667–1670

Kneipp K, Kneipp H, Itzkan I, Dasari RR, Feld MS (1999) Ultrasensitive chemical analysis by Raman spectroscopy. Chem Rev 99:2957–2975

Kneipp K, Kneipp H, Itzkan I, Dasari RR, Feld MS (2002) Surface-enhanced Raman scattering and biophysics. J Phys Condens Matter 14:597–624

Kruse PW, McGlauchlin LD, McQuistan LB (1962) Elements of infrared technology. Wiley & Sons, New York

Kramers HA, Heisenberg W (1925) Über die Streuung von Strahlung durch Atome. Zeitschrift für Physik 31:681–707

Leopold N, Cinta-Panzaru S, Baia M, Antonescu E, Cozar O, Kiefer W, Popp J (2005) Raman and surface-enhanced Raman study of thiamine at different pH values. Vib Spectrosc 39:169–176

Lommel E (1878) Theorie der Absorption und Fluorescence. Ann Phys u Chem 3:251–283

Maiman TH (1960) Stimulated optical radiation in ruby. Nature 187:493–494

Moskovits M (1985) Surface-enhanced spectroscopy. Rev Mod Phys 57:783–826

Moskovits M (1982) Surface selection rules. J Chem Phys 77:4408–4416

Moskovits M, Suh JS (1984) Surface selection rules for surface-enhanced Raman spectroscopy: calculations and application to the surface-enhanced Raman spectrum of phthalazine on silver. J Phys Chem 88:5526–5530

Nakamoto K (1997) Infrared and Raman spectra of inorganic and coordination chemistry, part A: theory and applications in inorganic chemistry. 5th edn. Wiley & Sons, New York

Nafie LA (2001) Theory of Raman scattering. In Lewis IR, Eduards HGM (eds) Handbook of Raman spectroscopy, Marcel Dekker Inc, New York

Nie S, Emory SR (1997) Probing single molecules and single nanoparticles by surface-enhanced Raman scattering. Science 275:1102–1106

Parr RG, Yang W (1989) Density functional theory of atoms and molecules, Oxford University Press, Oxford

Placzek G (1934) Rayleigh Streuung und Ramaneffekt. In: Marx G (ed) Handbuch der Radiologie, vol. 6, Akad. Verlagsgesellschaft, Leipzig

Raman CV, Krishnan KS (1928) A new type of secondary radiation. Nature 121:501–502

Riboh JC, Haes AJ, McFarland AD, Yonzon CR, Van Duyne RP (2003) A nanoscale optical biosensor: Real-time immunoassay in physiological buffer enabled by improved nanoparticle adhesion. J Phys Chem B 107:1772–1780

Rosi NL, Mirkin CA (2005) Nanostructures in biodiagnostics. Chem Rev 105:1547–1562

Schrödinger E (1926) Quantisierung als Eigenwertproblem IV. Ann d Physik 81:109–139

Sequaris JM, Koglin E (1985) Subnanogram colloid surface-enhanced Raman spectroscopy (SERS) of methylated guanine on silica gel plates. Anal Chem 321:758–759

Seminario JM, Politzer P (1995) Modern density functional theory: a tool for chemistry. Elsevier, Amsterdam

Smekal AG (1923) Zur Quantentheorie der Dispersion. Naturwissenschaften 11:873–875

Srituravanich W, Fang N, Sun C, Luo Q, Zhang X (2004) Plasmonic nanolithography. Nano Letters 4:1085–1088

Toderas F, Mihut A M, Baia M, Simon S, Astilean S (2004a) Self-assembled gold nanoparticles on solid substrate. Studia UBB Physica XLIX:89–94

Toderas F, Mihut A, Baia M, Astilean S (2004b) Self-assembling noble metal nanoparticles and their investigation for applications in surface-enhanced Raman spectroscopy (SERS) and surface plasmon resonance (SPR) biodetection. Book of Abstracts of the 1st International Conference Advanced Spectroscopies on Biomedical and Nanostructured Systems, Cluj-Napoca, 96

Toderas F, Boca S, Baia M, Baia L, Maniu D, Astilean S, Simon S (2006a) Self-assembled multilayers of gold nanoparticles as versatile platforms for molecular sensing by Fourier transform-surface enhanced scattering (FT-SERS) and surface enhanced infrared absorption (SEIRA). Book of Abstracts of the 2nd International Conference Advanced Spectroscopies on Biomedical and Nanostructured Systems, Cluj-Napoca, 121

Toderas F, Baia M, Baia L, Maniu D, Farcau C, Astilean S, Barbu-Tudoran L, Craciun C (2006b) Gold nanoparticles self-assembled on functionalized glass substrates and their surface plasmons enhanced properties. Book of Abstracts of the International Conference Micro to Nano-Photonics, ROMOPTO, Sibiu, 45

Toderas F, Baia M, Baia L, Astilean S (2007) Controlling gold nanoparticle assemblies for efficient surface enhanced Raman scattering (SERS) and localized surface plasmon resonance (LSPR) sensors. Nanotechnology, 18:doi:10.1088/0957-4484/18/25/255702

Van Duyne RP (2004) Molecular plasmonics. Science 306:985–986

Vo-Dinh T (1988) Surface-enhanced Raman spectroscopy using metallic nanostructures. Trends in Analyt Chem 17:557–582

2 Fundamentals of Infrared and Raman Spectroscopy, SERS, and Theoretical Simulations

Infrared and Raman spectroscopy are two important vibrational spectroscopy methods. These investigation techniques provide complementary images of molecular vibrations, because the mechanisms of the interaction of light with molecules in those two spectroscopic techniques are quite different.

2.1 Vibrations of Molecules

Molecules consist of atoms with a certain mass that can be considered as being connected by elastic bonds. As a result, they can perform periodic motions, and have vibrational degrees of freedom. All motions of the atoms in a molecule relative to each other are a superposition of so-called *normal vibrations*, in which all atoms are vibrating with the same phase and normal frequency. Their amplitudes are described by a normal coordinate. Polyatomic molecules with N atoms possess *3N-6* normal vibrations (linear ones have *3N-5* normal vibrations), which define their vibrational spectra (Demtröder 1981, Chalmers and Griffiths 2002, Hollas 1992, Bunker 1979, Niquist 2001). These spectra depend on the masses of the atoms, their geometrical arrangement, and the strength of their chemical bonds. Depending on whether the bond length or angle is changing, there are two types of molecular vibrations, stretching and bending. The stretching vibrations can be symmetric and asymmetric, whereas the bending vibrations can be subdivided in scissoring, rocking, wagging and twisting vibrations. For molecules with certain elements of symmetry, some vibrational modes may be degenerate, so that more than one mode has a given vibrational frequency, while others may be completely forbidden. Thus, because of degeneracy, the number of the observed fundamental absorption bands is often less than *3N-6*. In "super molecules," such as crystals or complexes, the vibrations of the individual components are coupled.

The simplest model of a vibrating molecule describes an atom bound to a very large mass by a weightless spring (Schrader 1995). The force F, which is necessary to move the atom by a certain distance x from an equilibrium position, is given by Hooke's law, and is proportional to the force constant f, which represents a measure of the bond strength:

$$F = -fx .$$
(2.1)

By using Newton's law, from where the force is also proportional to the mass M of the atom and its acceleration (the second derivate of the elongation with respect to the time, d^2x/dt^2), and solving the second order differential equation, which possesses the solution:

$$x = x_0 \cos(2\pi vt + \delta),$$
(2.2)

where δ represents the phase angle, the frequency of the vibration of the mass connected to a very large mass by an elastic spring can be expressed (Schrader 1995) as follows:

$$v = \frac{1}{2\pi}\sqrt{\frac{f}{M}} .$$
(2.3)

If one considers a diatomic molecule the mass M is the reduced mass of the diatomic molecule with the masses M_1 and M_2 and is given by the following formula (Schrader 1995):

$$\frac{1}{M} = \frac{1}{M_1} + \frac{1}{M_2} .$$
(2.4)

Equation 2.3 gives the frequency v (in Hz, s^{-1}) of the vibration. In vibrational spectroscopy, the wavenumber $\tilde{v}$ (in cm^{-1}) is frequently used instead of frequency, which is the reciprocal wavelength λ:

$$\tilde{v} = \frac{v}{c} = \frac{1}{\lambda} .$$
(2.5)

The use of dependences from Eq. 2.3 is the simplest way that can be employed in evaluating the effect of some structural changes on the band position in the case of vibrations, in which modifications of bonds length or atoms mass appear.

The potential energy of a molecule, which obeys Hooke's law, is obtained by integrating Eq. 2.1, and has the following expression (Schrader 1995):

$$V = \frac{1}{2} fR^2 ,$$
(2.6)

in which $R = x - x_c$, x_c represents the Cartesian coordinate of the potential minimum. Because the atoms vibrate with a definite frequency, according to the cosine function in Eq. 2.2, the potential energy given by Eq. 2.6 represents the harmonic potential of a molecule. In the complex structure of the substances, the molecular potentials are not exactly harmonic; therefore, an expression for an anharmonic potential was necessary (Schrader 1995). Thus, anharmonic molecular potentials have been represented by approximate functions, e. g., the Morse function (Morse 1929):

$$V = D\left(1 - e^{-R\sqrt{\frac{f_e}{2D}}}\right)^2,\tag{2.7}$$

where f_e is the force constant near the potential minimum and D stands for the dissociation energy. There are other empirical potential functions which are especially useful to describe intermolecular potentials, i. e., the potentials between atoms, which are not connected by chemical bonds (Kitaigorodski 1973, Pertsin and Kitaigorodski 1987). According to classical mechanics, a harmonic oscillator may vibrate with any amplitude, which means that it can possess any amount of energy, large or small. However, quantum mechanics show that molecules can only exist in definite energy states. In the case of harmonic potentials, these states are equidistant:

$$E_i = h\nu\left(\nu_i + \frac{1}{2}\right),\ \nu_i = 0, 1, 2, ...\tag{2.8}$$

while for anharmonic potentials, the distances between energy levels decrease with increasing energy. In Eq. 2.8 h and ν_i represent the Planck constant and the vibrational quantum number of each energy level, respectively. For $\nu_i = 0$ the potential energy has its lowest value, which is not the energy of the potential minimum. This is the so-called *zero point energy*. This energy cannot be removed from the molecule, even at temperatures approaching absolute zero (Schrader 1995).

For polyatomic molecules, the frequency of normal vibrations can be calculated by applying the Lagrange equation to expressions of the kinetic and potential energy of the molecule (Wilson et al. 1955). Generally, the potential function of a polyatomic molecule can be described by a Taylor series:

$$V(r) = V_0 + \sum_i f_i R_i + \frac{1}{2}\sum_{i,j} f_{ij} R_i R_j + \frac{1}{6}\sum_{i,j,k} f_{ijk} R_{ijk} + ...,\tag{2.9}$$

where R_i are suitable displacement coordinates from the equilibrium geometry at a potential minimum and the constants:

$$f_i = \left(\frac{\partial V}{\partial R_i}\right)_0,\ f_{ij} = \left(\frac{\partial^2 V}{\partial R_i \partial R_j}\right)_0,\ f_{ijk} = \left(\frac{\partial^3}{\partial R_i \partial R_j \partial R_k}\right)_0\tag{2.10}$$

are called linear, quadratic, cubic, etc. force constants. The potential can be defined such that $V_0 = 0$. By definition, the second term in Eq. 2.9 is equal to zero, because the molecule is regarded as being in its equilibrium (Califano 1976). Due to the fact that the cubic and higher terms have many components with a comparatively small value and are difficult to be determined, the potential function has the form (Schrader 1995):

$$V = \frac{1}{2}\left(\frac{\partial^2 V}{\partial R_i \partial R_j}\right)_0 R_i R_j = \frac{1}{2}\sum_{i,j} f_{ij} R_i R_j. \tag{2.11}$$

The diagonal force constants f_{ij} with $i = j$ describe the elasticity of a bond according to Hooke's law. The interaction force constants, the constants with $i \neq j$, describe the change of the elastic properties of one bond when another bond is deformed. Thus, the force constants of the bonds, the masses of the atoms, and the molecular geometry determine the frequencies and the relative motions of the atoms.

2.2 Infrared Spectroscopy

2.2.1 Basics

The infrared region of the electromagnetic spectrum extends from $14000\,\mathrm{cm}^{-1}$ to $10\,\mathrm{cm}^{-1}$. For chemical analysis, the region of most interest is the mid-infrared region ($4000\,\mathrm{cm}^{-1}$ to $400\,\mathrm{cm}^{-1}$) which corresponds to changes in vibrational energies within molecules. The far infrared region ($400\,\mathrm{cm}^{-1}$ to $10\,\mathrm{cm}^{-1}$) is useful for molecules containing heavy atoms such as inorganic compounds, but requires rather specialized experimental techniques. The infrared spectroscopic technique is a quick and relatively cheap technique, and it is useful for identifying certain functional groups in molecules. An infrared spectrum of a given compound is unique and can therefore serve as a fingerprint for this compound. However, it is rarely, if ever, possible to identify an unknown compound by using infrared spectroscopy alone.

Not all possible vibrations within a molecule will result in an absorption band in the infrared region. Interaction of infrared radiation with a vibrating molecule is possible if the electric vector of the radiation field oscillates with the same frequency as does the molecular dipole moment. A vibration is infrared active only if the molecular dipole moment μ is modulated by the normal vibration (Schrader 1995):

$$\left(\frac{\partial \mu}{\partial q}\right)_0 \neq 0, \tag{2.12}$$

where q describes the motion of the atoms during a normal vibration and the subscript 0 refers to the derivative taken at the equilibrium configuration. In other words, in order to be infrared active, the vibration must result in a change of dipole moment during the vibration. This means that for homonuclear diatomic molecules such as H_2, N_2, and O_2 no infrared absorption is observed, as these molecules have zero dipole moment and stretching of the bonds will not produce one. For heteronuclear diatomic molecules such CO and HCl, which possess a permanent dipole moment, infrared activity occurs because stretching of this bond leads to a change in dipole moment.

The recording of an infrared spectrum can be performed by using two ways of data acquisition. The first one uses a source of infrared radiation to produce a range of frequencies which are then separated into individual frequencies using a monochromator diffraction grating. The resulted beam is then split into two; one passes through the sample and the other one is used as a reference beam. The two beams then converge on the detector which measures the difference in intensity, the resulting plot representing a measure of transmission against frequency which is usually plotted as wavenumber (cm^{-1}). However, the slow scans using diffraction gratings are inefficient. Almost all modern infrared spectrometers use a different approach, the Fourier transform (FT) method, to scan the full spectral range at the same time. The central component of a FT spectrometer is a Michelson interferometer (Fig. 2.1). FT spectrometers operate by dividing the incoming radiation into two beams, subjecting each beam to a different time delay, and recombining the beams so that interference occurs (Barnes 1977).

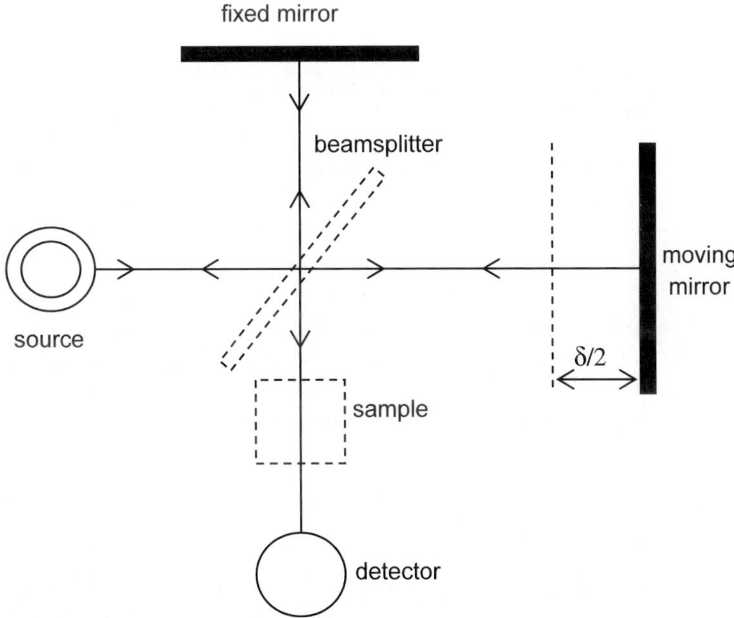

Fig. 2.1 The Michelson interferometer

The intensity I falling on the detector is a function of the optical path difference δ between the two beams (Bell 1972), which is produced by translating a mirror along the beams:

$$I(\delta) = \int_{0}^{\infty} P(\overline{v})\left[\frac{1}{2} + \cos(2\pi\overline{v}\delta)\right]d\overline{v} , \tag{2.13}$$

where $P(\overline{v})$ is the spectral power density at the wavenumber v.

Substrating the first term in the integral results in:

$$I'(\delta) = I(\delta) - \frac{1}{2}\int_{0}^{\infty} P(\overline{v})d\overline{v} = \int_{0}^{\infty} P(\overline{v})\cos(2\pi\overline{v}\delta)d\overline{v} . \tag{2.14}$$

which represents the total power.

A plot of $I(\delta)$, or $I'(\delta)$ against δ is known as an interferogram, which contains all the spectral information, but in a form that is not directly accessible. In order to obtain a spectrum it is necessary to calculate the Fourier transform of Eq. 2.14 (Barnes 1977).

$$P(\overline{v}) \propto \int_{-\infty}^{+\infty} I'(\delta)\cos(2\pi\overline{v}\delta)d\delta . \tag{2.15}$$

In practice, it is possible to record the interferogram over a restricted range of the path difference, and therefore the transform becomes:

$$P(\overline{v}) \propto \int_{-\delta_{\min}}^{+\delta_{\max}} I'(\delta)\cos(2\pi\overline{v}\delta)d\delta . \tag{2.16}$$

As a consequence, the bands become broader, leading to a lower resolution.

A practical spectrometer has to record data at finite sampling intervals. This may be conveniently achieved by moving the mirror over successive fixed distances, giving an optical path increment $\Delta\delta$. The effect of the finite sampling interval is to reduce the range of wavenumbers for which meaningful spectral information is obtained. It determines the maximum stepping interval that can be used in a particular wavenumber range. An alternative to stepping the moving mirror is to scan it rapidly at constant velocity v. Since the optical path difference is given by $\delta = 2vt$ Eq. 2.14 becomes (Barnes 1977):

$$I'(t) = \int_{0}^{+\infty} P(\overline{v})\cos(4\pi\overline{v}vt)d\overline{v} . \tag{2.17}$$

The total time for a given scan is determined by the maximum optical path difference D. The Fourier transform for a single-sided operation is then:

$$P(\overline{v}) \propto 2 \int_0^{D/2v} I'(t)\cos(4\pi\overline{v}vt)dt . \tag{2.18}$$

Weak spectral signals in rapid scan FT spectroscopy require that an interferogram be recorded repeatedly and the results (either in interferograms or the computed spectra) averaged to achieve the required signal-to-noise ratio.

An interferometer, or FT spectrometer, can offer certain inherent advantages over a conventional dispersive spectrometer:

1. Multiplex advantage (Fellgett's advantage): An interferometer provides information about the entire spectral range during the entire period of the measurement, whereas a dispersive spectrometer provides information only about the narrow wavenumber region which falls within the exit slit of the monochromator at any given time.
2. Throughput advantage (Jacquinot's advantage): The interferometer can operate with a large circular aperture, and using large solid angles at the source and at the detector, whereas a dispersive spectrometer requires long, narrow slits to achieve adequate resolution.

A few additional advantages follow from the multiplex and throughput advantages:

- Large resolution power: because the resolving power mainly depends on the maximum optical path difference, introducing high resolving power can be obtained by using large mirror movements. Also, unlike dispersive spectrometers, the wavenumber resolution is constant over the spectral range scanned.
- High wavenumber accuracy: the wavenumber accuracy is determined by the precision with which the position of the moving mirror can be measured.
- Fast scan time and a large wavenumber range are possible.

There are also some disadvantages of using an FT spectrometer over a conventional dispersive spectrometer. One of these disadvantages is that FT-infrared spectroscopy is normally a single beam technique, thus a comparison of the sample and the reference must always be performed by computer substraction.

2.2.2 FT-Infrared Measurements

The infrared spectra presented in the studies described in this book were recorded in the range from 400 to 4000 cm^{-1} with a BRUKER IFS 25 spectrometer. For infrared measurements, the samples were mixed with KBr and pressed by 9.8 Kbar in order to obtain thin pellets with a thickness of about 3 mm. The spectra were obtained with a spectral resolution of 2 cm^{-1}.

2.3 Raman Spectroscopy

2.3.1 Basics

When a molecule is exposed to an electric field, electrons and nuclei are forced to move in opposite directions and a dipole moment proportional to the electric field strength and to the molecular polarizability α is induced. A molecular vibration can only be observed in the Raman spectrum if there is a modulation of the molecular polarizability by the vibration (Schrader 1995):

$$\left(\frac{\partial \alpha}{\partial q}\right)_0 \neq 0 \, , \tag{2.19}$$

where q stands for the normal coordinates describing the motion of the atoms during a normal vibration, and the subscript 0 indicates that the derivative is taken at the equilibrium configuration.

The origin of Raman spectroscopy is an inelastic scattering effect, but in a Raman experiment the elastic as well as the inelastic scattering of radiation by the sample is observed.

The elastic scattering, which is also called Rayleigh scattering, corresponds to the light scattered at the frequency of the incident radiation ν_0. The molecule "absorbs" no energy from the incident radiation in this case. The inelastic scattered light, which is known as the Raman radiation, is shifted in frequency, and hence energy, from the frequency of the incident radiation by the vibrational energy that is gained or lost in the molecule $(h\nu_0 \pm h\nu_s)$ (see Fig. 2.2).

The Raman scattering process has a much lower probability in comparison with the Rayleigh scattering that is a coherent process. On the other hand, according to Boltzmann's law, most molecules are in their vibrational ground state at ambient temperature, a much smaller number being in the vibrationally excited state. Therefore, the Raman process, which transfers vibrational energy to the molecule and leaves a quantum of lower energy $(h\nu_0 - h\nu_s)$ has a higher probability than the reverse process, and the corresponding Raman lines are referred to as Stokes and anti-Stokes lines, respectively. The intensities of Stokes lines, caused by quanta of lower energy, are higher than those of anti-Stokes lines. Therefore, usually only Stokes radiation is recorded as a Raman spectrum (Schrader 1995).

In order to explain the Raman scattering process, virtual states have to be considered. This is related to the fact that the interaction of the photon with the molecule and the re-emission of the scattered photon occur almost simultaneously. The existence of such virtual states also explains why the non-resonance Raman effect does not depend on the wavelength of the excitation, since no real states are involved in this interaction mechanism. In fact, the Raman spectrum generally does not depend on the laser excitation.

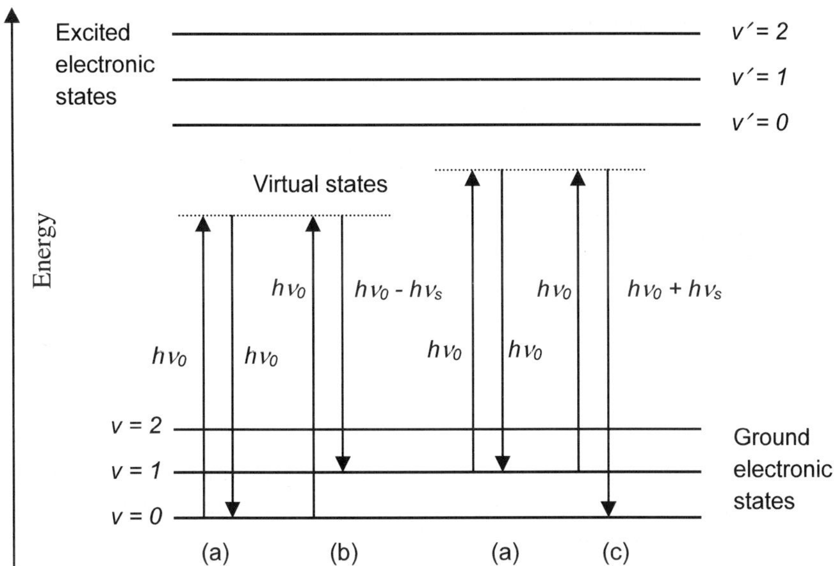

Fig. 2.2 Mechanisms of various light-scattering processes Rayleigh (*a*), non-resonance Stokes Raman (*b*), and non-resonance anti-Stokes Raman (*c*)

However, one of the most important aspects for using Raman spectroscopy in different application fields is the appropriate choice of the excitation wavelength of the laser. For analytical linear Raman spectroscopy, mostly continuous wave (CW) lasers with a fixed wavelength are applied. A large variety of lasers covering a broad range from the visible domain have been used in the last decades, but, due to the fast development of sensitive micro-Raman setups at the present time, red diode lasers operating at 785, 810, or 830 nm as well as cw Nd:YAG lasers (1064 nm) have been used in Raman spectroscopy. In particular, the development of those near-infrared lasers, which often avoid the excitation of Raman-masking fluorescence, has stimulated the field of biospectroscopy (Petry et al. 2003). Two different Raman spectrometers are usually applied for Raman spectroscopy, either a FT spectrometer or a dispersive one.

Twenty years ago Hirschfeld and Chase (Hirschfeld and Chase 1986) demonstrated that Raman spectra, excited with lasers in the near-infrared region, may be recorded with the FT-infrared instruments originally designed for absorption spectroscopy in the near-infrared region. They demonstrated that the excitation of fluorescence, very often associated with Raman spectroscopy excited in the UV-vis region, can be avoided. Raman spectra of fluorescing dyes and even of explosives were obtained, demonstrating that the thermal load by the illumination with the laser was not too large. They proved that the prejudice of many scientists against this technique (Hirschfeld 1976, Hirschfeld and Schildkraut 1974) was apparently not justified. Immediately after this publication, several groups started to apply this technique and instrument manufacturers began designing FT-Raman spectrometers.

During this time special attention was focused on developing special techniques able to ensure the samples investigations with a small spatial resolution. Thus, nowadays the micro-Raman technique, in which a microscope objective is employed and serves to focus the laser beam, is a well-established method for the investigation of samples in the order of picograms or even less (Petry et al. 2003). Microscope objectives with a high numerical aperture and high magnification are used to focus the light down to the diffraction limit. The FT-Raman spectroscopy with long-wavelength excitation (cw Nd:YAG 1064 nm laser) is also available in combination with a microscope.

2.3.2 FT-Raman Measurements

All FT-Raman spectra discussed in the next chapters were recorded at room temperature using a BRUKER IFS 120HR spectrometer equipped with a FRA 106 Raman module. The spectral resolution was $2\,\mathrm{cm}^{-1}$. Radiation of 1064 nm from a Nd-YAG laser with an output power comprised between 200 and 800 mW was employed for excitation. A Ge detector cooled with liquid nitrogen was used.

2.4 Surface-Enhanced Raman Spectroscopy

2.4.1 Basics

The SERS effect consists of the enormous enhancement of the Raman signal of the molecules adsorbed on roughened metallic surfaces (Fleischmann et al. 1974, Jeanmaire and Van Duyne 1977). The enhancement factors for the Raman scattering signals of adsorbed molecules were found to be more than a millionfold in comparison with the normal Raman signals (Albrecht and Creighton 1977). Many of the advantages of Raman scattering, such as molecularly specific vibrational spectra, simple versatile sampling, and the ready determination of analytes in air under vacuum and in water are applicable to this technique. However, with SERS, increased sensitivity is obtained and much lower concentrations can be studied. Detection limits are considerably lower (down to $10^{-9}\,\mathrm{M}$) than those for resonance Raman scattering (Sequaris and Koglin 1985).

A qualitative understanding of the SERS process is provided by the classical theory of the light scattering (Vo-Dihn 1998). One considers an incident light beam that induces an oscillation dipole μ in a particle which reemits or scatters light at the same frequency of the dipole oscillation. For the particular case where the magnitude of the incident electric field E is not too large, the induced dipole moment can be approximated as follows:

$$\mu = \alpha E \qquad (2.20)$$

where α is the polarizability of the molecule.

Having in mind that the Raman intensity is proportional to the square of the induced dipole μ one can assume that there are two possible enhancement mechanisms. Thus, the enhancement effect can influence either the molecular polarizability or the electric field experienced by the molecule. Surface selection rules are also available (Moskovits 1982, Creighton 1988). According to them, the most intense bands are those given by vibrations, which induce a polarization of the adsorbate electron cloud perpendicular to the metal surface (Creighton 1988, Moskovits and Suh 1984). By using the surface selection rules the orientation of the adsorbed molecule relative to the metal surface can be predicted (Gao et al. 1990, Moskovits and Suh 1984).

Despite the difficulties and limitations of SERS, i. e., poor quantitative reproducibility and the surface photolysis effect, the technique has become an increasingly popular analytical tool, which has been applied in numerous fields such as biomedicine, biophysics and biochemistry, surface science, analytical and environmental applications (Cotton et al. 1991, Baker and Moore 2005, Haynes et al. 2005, Dieringer et al. 2006, Rosi and Mirkin 2005, Kneipp et al. 2002, Kim et al. 2007), and so on.

2.4.2 Mechanisms of Surface Enhancement

The nature of the mechanisms, which produce the surface enhancement effect, is still not totally elucidated and remains the focus of debate (Weitz et al. 1986, Otto et al. 1992). As was mentioned above, there are two major types of contributions to the total enhancement of the Raman signal of adsorbed molecules: an electromagnetic enhancement sometimes referred to as the electric effect in which the molecule experiences large local fields caused by electromagnetic resonances occurring near metal surface structures, and a chemical or charge-transfer effect also referred to as the molecular effect in which the molecular polarizability is affected by the interaction between the molecule and the metal surface (Vo-Dinh 1998). Most researchers believe that much of the enhancement is due to the electromagnetic mechanism (Weitz et al. 1986, Otto et al. 1992). However, the significant evidence of the contribution of the charge transfer mechanism to the overall SERS enhancement cannot be ignored (Otto 1991, Guzonas et al. 1990). Discussions over the relative contribution to the total enhancement of both these mechanisms are ongoing.

2.4.2.1 Electromagnetic Enhancement

The electromagnetic enhancement is based on two main enhancements. The first one is the enhancement of the laser electromagnetic field due to the addition of the field provoked by the polarization of the metal particles. The second one is due to the molecule radiating an amplified Raman field, which further polarizes the metal particle and in this manner acts as an antenna to further amplify the Raman signal.

Surface roughness is an essential requirement of SERS. On a smooth metal surface, surface plasmons exist as waves of electrons bound to the metal surface and are capable of moving only in a direction parallel to the surface. On a roughened metal surface, the plasmons are no longer confined and the resulting electric field can radiate both in a parallel and a perpendicular direction to the surface. When an incident photon falls on the roughened surface, excitation of the plasmon resonance of the metal may occur and this allows scattering. Additionally, due to the difference in dielectric constants between the roughened surface and the surrounding media, a concentration of electric field density occurs at sharp points on the surface (Weitz et al. 1986, Otto et al. 1992). Metal colloids and colloidal aggregates provide a particularly rich example of such local electromagnetic enhancement. Several authors (Xu et al. 2000, Gesten and Nitzan 1980, Gesten and Nitzan 1985, Inoue and Ohtaka 1989) suggest that protrusions on the surface of a colloidal particles, as well as "cavities" between adjacent particles in an aggregate, lead to a giant enhancement of the local field, up to a factor of 10^{14}–10^{15}.

2.4.2.2 The Charge Transfer Mechanism

Numerous studies have been carried out in order to elucidate the existence of the charge transfer mechanism (Otto 1991, Guzonas et al. 1990). Some researchers use it to explain why the enhancement factor of the first adsorbate layer is much greater than that of the subsequent layers. Basically, the enhancement experienced from the charge transfer results when molecules physisorb or chemisorb directly on the roughened surface, forming an adsorbate-metal complex. If chemisorption occurs, the molecular orbitals of the adsorbate are broadened by an interaction with the conduction bands of the metal surface. This results in a ready transfer of electrons and excitation from the metal to the adsorbate and vice versa. As a consequence, the SERS spectra of chemisorbed molecules are significantly different from the Raman spectrum of the free species, although those of the physisorbed species are not changed. The charge transfer mechanism is restricted by its nature to molecules directly adsorbed on the metal, as opposed to the electromagnetic effect, which extends a certain distance beyond the surface. Thus, it effectively operates only on the first layer of adsorbates.

Campion and coworkers (Campion et al. 1995) reported the first experimental evidence of the charge transfer mechanism linking new features in the electronic spectrum of an adsorbate to SERS, under conditions where electromagnetic enhancements are unimportant. They stated that it was difficult to observe only the charge transfer because electromagnetic effects had to be accounted for and removed. However, they overcame this problem by measuring SERS enhancement on a flat, smooth single crystal surface where electromagnetic effects were small and well understood. Hildebrandt and Stockburger (Hildebrandt and Stockburger 1984) carried out an extensive study of surface enhanced resonance Raman scattering (SERRS) of rhodamine 6G on colloids, to explore the enhancement mechanisms involved in this technique. They reported that two different types of adsorption sites from the colloid surface were responsible for the observed enhancement: a non-specific adsorption site that had a high surface coverage on the colloid surface, which resulted in an enhancement factor of 3000 and could be explained by a classical electromagnetic mechanism, and a specific adsorption site that was activated only in the presence of certain anions (Cl^-, I^-, F^-, Br^-, and SO^{4-}). Although this specific site had a low surface coverage (approximately three per colloidal particle) the authors claimed an enhancement of 10^6, which was believed to be due to a charge transfer mechanism. In a further study (Hildebrandt and Stockburger 1986), they concluded that charge transfer enhancement is strongly dependent upon the structural and electronic properties of the analyte.

The understanding and experimental proof supporting this enhancement mechanism is limited. The problem is even more complex due to the fact that electromagnetic enhancement increases as the adsorbate-surface distance decreases, and only the additional enhancement can be classified as charge transfer. However, the degree of enhancement of the first layer is very large. Thus, many questions remain unanswered, and therefore, the charge transfer mechanism is not yet completely accepted.

2.4.3 SERS-Active Substrates

Surface enhancement is observed from a limited number of roughened metals, i. e., silver, gold, copper, aluminum, lithium, and sodium. The intensity of scattering from adsorbed analytes is no longer proportional to the frequency to the fourth power (v^4), as in the case of the conventional Raman process; in fact, the intensity of the bands is related to the frequency of the surface plasmon resonance and the laser excitation frequency (Fleischmann et al. 1974, Jeanmaire and Van Duyne 1977, Albrecht and Creighton 1977). The exact dependence is related to the nature of the metal substrate, in particular to the identity of the metal and its roughness. Since SERS was first observed, numerous SERS-active substrates have been developed.

Several different metal electrodes have been employed for SERS, but the largest surface enhancement was observed from those made of silver (Fleischmann et al. 1974). When using an electrode as a SERS substrate, the surface roughness can be controlled by the right choice of electrolytes and electrochemical cycle. The degree of adsorption is also affected by the applied electrode potential during the Raman measurement that modifies the metal Fermi level (Vo-Dinh et al. 1999). The main disadvantage of this substrate is the reproducibility of the electrode surface; it is incredibly difficult to ensure that the same degree of surface roughness and potential is achieved for each experiment.

Colloidal suspensions are attractive as SERS substrates, since the aggregation of metal particles leads to the formation of aggregates with the roughness and fractal morphology necessary to render intense Raman spectra (Albrecht and Creighton 1977, Sanchez-Gil and García-Ramos 1998) and they can be prepared with a high reproducibility. Furthermore, they are relatively inexpensive. Because a fresh, reproducible colloidal surface is available for each analysis, reliable SERS analysis is possible. Numerous metal colloidal suspensions have been used, including gold and copper; however, silver is the most popular one.

Among the methods employed to obtain metal colloids, the chemical reduction of silver nitrate with citrate (Lee and Meisel 1982) produces a more uniform distribution of particle sizes (24–30 nm). It was found (Rivas et al. 2001) that the SERS intensity increases as the size of the silver particles becomes larger (average diameter about 50 nm). However, a further increase of the particle size (100–130 nm) leads to a lower SERS intensity signal. Thus, if the nanostructure responsible for SERS becomes too small, the effective conductivity of the metal nanoparticles diminishes as a result of electronic scattering processes at the particle's surface and consequently, the quality factor of the dipolar plasmon resonance is vitiated and the re-radiated field strength is reduced (Moskovits 2005). On the other hand, the upper dimensional bound of the SERS-active system is determined by wavelength. When nanoparticles of the order of the wavelength or larger are used, the optical fields excite progressively higher order multipoles that are non-radiative and hence are not efficient in exciting Raman excitations (Moskovits 2005). Thus, one of the advantages of metal colloids is the possibility to control and modify the particle size and shape by choosing adequate experimental conditions (Ahmadi et al. 1996).

The silver colloid is stabilized against coagulation by electrostatic forces originating from anions adsorbed on the particle surface, which are citrate anions in the case of the silver nitrate reduction with sodium citrate. In many cases, these adsorbed anions disturb the adsorption of analyte molecules. To avoid this perturbation, a stable and reproducible colloid was prepared by reduction of the silver nitrate with hydrazine (Nikel et al. 2000). At alkaline pH values (around 10) the stabilizing agent is most likely OH^-, which is less tightly bonded than, for example, citrate anions in the colloid prepared according to the procedure of Lee and Meisel (Lee and Meisel 1982). A small amount of strongly adsorbing cationic analyte molecule can completely displace OH^- and concomitantly establish a positive surface charge that, in turn, electrostatically stabilizes the colloid.

Murphy et al. (Murphy et al. 1999) prepared a SERS-active substrate, which consists of silver or gold colloids encapsulated in a sol-gel derived xerogel layer. Control of the gel parameters, such as porosity, pore size, and polarity, enables the tailoring of sensitivity to different analyte groups.

Some microorganisms, such as bacteria, can produce metal colloids by mechanisms, which are not yet fully understood (Zeiri et al. 2002). One possible role of the bacteria is in providing a multitude of nucleation centers, establishing conditions for obtaining highly dispersed, small nanoparticle systems. In addition, they slow down aggregation, or entirely prevent it, by immobilizing the particles and providing a viscous medium. One possible application of the silver colloid deposits in bacteria is the obtaining of intense SERS spectra (Efrima 1985) to probe the immediate biochemical environment near the metal cores. SERS spectra obtained in this way are particularly sensitive to one specific component from the cell wall, that is riboflavin (or flavinadenine dinucleotide), which is a co-factor of major importance in a variety of live-sustaining processes in living cell.

Recently, there has been a renewed interest in the enhancement of isolated nanoparticles deposited on planar surfaces (Emory and Nie 1997, Vlckova et al. 1996). Oldenberg et al. (Oldenberg et al. 1999) obtained a high enhancement factor in SERS by using gold nanoshell as a support. A gold nanoshell is a composite nanoparticle consisting of a dielectric core coated by a thin metal shell, its peak plasmon resonance wavelength being determined by the ratio of the core diameter to the shell thickness.

The growth of the metal particles obtained by citrate reduction of the silver nitrate (Turkevich et al. 1951) is strongly temperature dependent, and the preparation is typically performed in boiling water. The simple mixing of a standard solution of silver salt and citrate at room temperature does not result in a detectable nanoparticle formation. Silver particles are formed at room temperature with a spatial control on the micrometer scale, when the solution in contact with the standard cover glass is illuminated with a tightly focused low-power laser beam (Bjeneld et al. 2002).

A great variety of surface-confined nanostructures, that allow the control of the plasmonic properties and the improvement of the Raman signal enhancement, can be produced by numerous fabrication methods, including colloid immobilization, electron-beam lithography, and nanosphere lithography.

Based on the consideration that large enhancement factors can be obtained at the junctions of aggregated nanoparticles, colloidal nanoparticles of different sizes and shapes were immobilized on functionalized solid substrates and were found to be highly efficient SERS substrates (Baia et al. 2006d, Li et al. 2004, Wang and Gu 2005, Orendorff et al. 2005). Moreover, by fabricating the junctions of the metal nanoparticle-molecule-metal nanoparticle on glass in sandwich architectures, large enhancements were also obtained at the junctions (Hu et al. 2007, Li et al. 2007, Wei et al. 2007, Zhou et al. 2007, Zhou et al. 2006).

Lithography techniques can be exploited to fabricate topographically predictable SERS substrates. Electron-beam lithography allows the fabrication of nanostructures of desired shape, size, and arrangement. The substrates obtained by this technique were used to explore how the magnitude of the enhancement factor is influenced, when the size, shape, and the interparticle spacing is varied (Gunnarsson et al. 2001, Grand et al. 2005, Grand et al. 2003, Felidj et al. 2004, Felidj et al. 2002). In the case of nanosphere lithography, a colloidal crystal monolayer of size-monodisperse nanospheres is grown on a flat substrate, and then the desired nanoparticle material is thermally evaporated through the nanosphere mask. The metallic nanostructured films deposited over polystyrene crystal templates have SERS activity (Haynes et al. 2005, Baia et al. 2005, Baia et al. 2006a, Astilean et al. 2004, Astilean et al. 2005). Upon removing the nanospheres from the surface, (Haynes and Van Duyne 2002, Haynes and Van Duyne 2003) homogenous arrays of truncated tetrahedral nanoparticles, that are also highly efficient SERS-active substrates (Schmidt et al. 2004, McFarland et al. 2005, Baia et al. 2006b, Baia et al. 2006c, Astilean et al. 2006), remain on the surface. These SERS-substrates have controllable SERS-activity depending on the film thickness and the dimensions of the polystyrene nanospheres employed for obtaining the deposition mask.

Researchers continue to develop novel SERS substrates to prolong substrate lifetime, to provide stable and optimized enhancement factors, and to permit SERS studies in diverse environments.

2.4.4 SERS Measurements

In the majority of the SERS studies presented in the following chapters, a sodium citrate silver colloid, prepared according to the standard procedure of Lee and Meisel (Lee and Meisel 1982), was employed as a SERS substrate. $AgNO_3$ (90 mg) was dissolved in 500 ml of water and heated to boiling with continuous stirring. A 10 ml portion of 1% aqueous trisodium citrate was added drop wise, and the reaction mixture was boiled for another 60 min. The resultant colloid was yellowish gray with an absorption maximum at 407 nm. NaCl solution (10^{-2} M) was added (10:1) to produce an aggregation of the colloidal dispersion that yields to a considerable enhancement of the SERS signal (Brandt and Cotton 1993). The final concentration of the samples into colloidal suspension was approximately $3 \cdot 10^{-4}$ M. NaOH, HCl, and H_2SO_4 were used in order to adjust the pH values. All starting materials involved in substrate and sample preparation were purchased from commercial sources as analytical pure reagents.

The gold colloidal suspension was prepared by the following procedure: 500 ml of 10^{-3} M $HAuCl_4$ was brought to a boil with vigorous stirring on a magnetic stirring hot plate. Ten milliliters of 38.8 mM Na_3 citrate was added to the solution all at once with vigorous stirring. The yellow solution turned clear, then dark blue, and then a deep red/burgundy color within a few minutes. Stirring and boiling was continued for 10–15 min after the burgundy color was observed. The solution was removed from heat and continuously stirred until it got cold, and then the volume was adjusted to 500 ml with water. Colloidal solutions were stored in clean brown glass bottles until used.

The SERS spectra of the samples in silver colloid were collected in a 180° back-scattering arrangement. For excitation the 514.5 nm line (300 mW) of a Spectra Physics argon ion laser was used. The scattered Raman light was analyzed with a Spex 1404 double monochromator and the dispersed Raman stray light was detected with a Photometrics model 9000 CCD camera. The spectral resolution was $2 \, cm^{-1}$.

The SERS measurements of the samples in the gold colloid were performed in back-scattering geometry with a Dilor Labram system equipped with a microscope objective, a 950 lines/mm grating, and an internal HeNe laser with an emission wavelength of 632.81 nm. The spectral resolution was of $4 \, cm^{-1}$.

2.5 Theoretical Simulations

Computational methods simulate chemical structures and reactions numerically, based totally or partially on the fundamental laws of physics. Some methods can be used to model not only stable molecules, but also short-lived, unstable intermediates, and even transition states. In this way, they can provide information about molecules and reactions, which is extremely difficult to be obtained experimentally. Therefore, computational methods represent both an independent research area and a vital adjunct to experimental studies (Foresman 1996).

2.5.1 Molecular Mechanics and Electronic Structure Methods

There are two broad areas within computational chemistry (Szabo and Ostlund 1982) dealing with the structure of molecules and their reactivity: molecular mechanics and electronic structure theory.

Molecular mechanics simulations use the laws of classical physics to predict the structures and the properties of the molecules. There are many different molecular mechanics methods, each one being characterized by its particular force field. These methods do not explicitly treat the electron in a molecular system. The calculations are performed based upon the interactions among the nuclei, and the electronic effects are implicitly included in force fields through parametrization. Therefore, molecular mechanics simulations are quite inexpensive computationally and can be used for very large systems containing thousands of atoms. However, they also have limitations: no force field can be usually used for all molecular systems of interest and they cannot describe molecular properties, which depend on subtle electronic details (Szabo and Ostlund 1982).

On the other hand, electronic structure methods use the laws of quantum mechanics rather than of classical physics as the basis for their computations. According to quantum mechanics, the energy and other related properties of a particle might be obtained by solving the Schrödinger equation:

$$\left\{ \frac{-h^2}{8\pi^2 m} \nabla^2 + V \right\} \psi(\vec{r},t) = \frac{ih}{2\pi} \frac{\partial \Psi(\vec{r},t)}{\partial t}. \tag{2.21}$$

In this equation Ψ is the wavefunction that depends on the coordinates $\vec{r}$ and time t, m is the mass of the particle, h is Planck's constant, and V is the potential field in which the particle is moving. For a collection of particles like a molecule the Schrödinger equation is very similar. In this case, Ψ would be a function of the coordinates of all particles in the system as well as of the time t. However, exact solutions of the Schrödinger equation are not computationally practical (Szabo and Ostlund 1982).

Electronic structure methods are characterized by their various mathematical approximations used to solve the Schrödinger equation. Thus, there are two major classes of electronic structure methods:

Semi-empirical methods (AM1, PM3) that use parameters derived from experimental data to simplify the computation (Dewar and Reynolds 1986, Stewart 1989). They solve an approximate form of the Schrödinger equation that depends on having appropriate parameters available for the type of chemical system under investigation. Different *semi-empirical* methods are largely characterized by their different parameter sets.

Ab initio methods (Hehre et al. 1986) that, unlike either molecular mechanics or *semi-empirical* methods, use no experimental parameters. Their computations are based solely on the laws of quantum mechanics and on the values of a small number of physical constants: the speed of light, the masses and charges of electrons and nuclei, and Planck's constant. *Ab initio* methods use a series of rigorous mathematical approximations to solve the Schrödinger equation.

Semi-empirical and *ab initio* methods differ quite a lot between the computational cost and the accuracy of the results. *Semi-empirical* methods are relatively inexpensive and provide reasonably qualitative descriptions of molecular systems and fairly accurate quantitative predictions of energies and structures for systems where good parameter sets exist. In contrast, *ab initio* computations provide high quality quantitative predictions for a broad range of systems. They are not limited to any specific class or size of a system (Hehre et al. 1986).

In the last years a third class of electronic structure methods has gained steadily in popularity: *density functional* methods (Frisch et al. 1998). These density functional theory (DFT) methods are similar to *ab initio* in many ways. DFT calculations require about the same amount of computation resources as the Hartree–Fock (HF) theory, the least expensive *ab initio* method. DFT methods are attractive because they include in their model the effects of electron correlation, the fact that electrons in a molecular system react to one another's motion and attempt to keep out of one another's way. HF calculations consider this effect only in an average sense – each electron sees and reacts to an averaged electron density – while methods including electron correlation account for the instantaneous interactions of pairs of electrons with opposite spin. This approximation causes HF results to be less accurate for some types of systems. Thus, DFT methods can provide the benefits of some more expensive *ab initio* methods at essentially HF cost.

2.5.2 The Simulation Model

A simulation model has been defined as an unbiased, uniquely defined, and uniformly applicable theoretical model for predicting the properties of chemical systems (Foresman 1996). It generally consists of the combination of a theoretical method with a basis set. Each such pairing of a method with a basis set represents a different approximation of the Schrödinger equation. Other desirable features of a simulation model include the following:

Size consistency: the results given for a system of molecules infinitely separated from one another ought to equal the sum of the results obtained for each individual molecule calculated separately. Another way of describing this requirement is that the error in the predictions of any method should scale roughly in proportion to the size of the molecule. When size consistency is not respected, the comparison of the properties of molecules of different sizes will not give qualitatively meaningful differences.

Reproducing the exact solution for the relevant n-electron problem: a method has to yield the same results as the exact solution of the Schrödinger equation to the greatest possible extent. What this means specifically depends on the theory underlying the method.

Variational: the energies predicted by a method ought to be an upper bound to the real energy resulting from the exact solution of the Schrödinger equation.

Efficient: calculations with a method must be practical with the existing computer technology.

Accurate: ideally, a method has to produce highly accurate quantitative results. A method should at least predict qualitative trends for molecular properties for groups of molecular systems.

The results obtained for different chemical systems generally may be compared only if they have been predicted via the same simulation model. Different models could be compared and tested by comparing their results for the same systems and with the results of experiments (Foresman 1996).

2.5.3 *DFT Methods*

DFT methods compute electron correlation via general *functionals* of the electron density. Such methods owe their modern origin to the Hohenberg–Kohn theorem (Hohenberg and Kohn 1964), which demonstrates the existence of a unique functional, which determines exactly the ground state energy and density. However, the theorem does not provide the form of this functional. The approximate functionals employed by current DFT methods divide the electronic energy into several terms:

$$E = E_k + E_p + E_{e-e} + E_{ex} \qquad (2.22)$$

where E_k is the kinetic energy term (arising from the motion of the electrons), E_p includes terms describing the potential energy of the nuclear-electron attraction and of the repulsion between pairs of nuclei, E_{e-e} is the electron-electron repulsion term (it is also described as the Coulomb self-interaction of the electron density), and E_{ex} is the exchange-correlation term and includes the remaining part of the electron-electron interactions. All terms except the nuclear-nuclear repulsion are functions of the electron density ρ. The sum $E_k + E_p + E_{ex}$ corresponds to the classical energy of the charge distribution, whereas the E_{ex} term accounts for the exchange energy arising from the antisymmetry of the quantum mechanical wavefunction and the dynamic correlation in the motions of the individual electrons. Hohenberg and Kohn (Hohenberg and Kohn 1964) demonstrated that E_{ex} is completely determined by the electron density ρ and is usually divided into separate parts, referred to as the exchange and correlation parts, but corresponds to the same-spin and mixed-spin interactions, respectively:

$$E_{ex}(\rho) = E_{ex-e}(\rho) + E_{ex-c}(\rho). \tag{2.23}$$

These terms are again functionals of the electron density, and the functional defining the two components on the right side of Eq. 2.23 are termed *exchange functionals* E_{ex-e} and *correlation functionals* E_{ex-c}, respectively. Both components can be of two distinct types: *local* functionals and *gradient corrected* functionals.

Local exchange and *correlation* functionals depend only on the value of the electron spin densities. Slater and $X\alpha$ are well-known local exchange functionals (Slater 1974), whereas the local spin density treatment of Vosko, Wilk, and Nusair (VWN) is a widely used local correlation functional (Vosko et al. 1980).

Gradient-corrected functionals involve both the values of the electron spin densities ρ and their gradients $\nabla\rho$. Such functionals are sometimes referred in the literature as *non-local*. A popular gradient-corrected exchange functional is the one proposed by Becke in 1988 (Becke 1988), while a widely-used gradient-corrected correlation functional is the LYP functional of Lee, Yang, and Parr (Lee et al. 1988). The combination of the two forms is referred as the BLYP method. Perdew has also proposed some important gradient-corrected correlation functionals, known as Perdew 86 and Perdew-Wang 91 (Becke 1993).

There are also several *hybrid* functionals, which define the exchange functional as a linear combination of Hartree–Fock, local, and gradient-corrected exchange terms; this exchange functional is then combined with a local and/or gradient-corrected correlation functional. The best known hybrid functional is Becke's three-parameter formulation (Lee et al. 1988, Becke 1993, Perdew and Wang 1992); hybrid functionals based on it are available in Gaussian (Frisch et al. 1998) via the B3LYP and B3PW91 keywords. Becke-style hybrid functionals have proven to be superior to the traditional functionals defined so far.

2.5.4 The Basis Set

A *basis set* is a mathematical description of the orbitals within a system used to perform the theoretical calculation (Szabo and Ostlund 1982). The basis set can be interpreted as restricting each electron to a particular region of space. Larger basis sets impose fewer constrains on electrons and approximate each orbital more accurately, but require more computational resources. Standard basis sets for electronic structure calculations use linear combinations of basis functions (one-electron functions) to form the orbitals. An individual molecular orbital is defined as follows (Szabo and Ostlund 1982):

$$\phi_i = \sum_{\mu=1}^{N} c_{\mu i} \chi_\mu \tag{2.24}$$

where the coefficients $c_{\mu i}$ are known as the molecular orbital expansion coefficients, and the basis functions $\chi_1 \dots \chi_N$ are chosen to be normalized.

Gaussian and other *ab initio* electronic structure programs use basis functions which are themselves composed of a linear combination of Gaussian functions; such basis functions are referred to as contracted functions, and the component Gaussian functions are referred as primitives. A basis function consisting of a single Gaussian function is termed uncontracted. The Gaussian program package (Frisch et al. 1998) offers a wide range of predefined basis sets, which may be classified by the number and type of basis functions that they contain as follows.

Minimal basis sets contain the minimum number of basis functions needed for each atom. They use fixed-size atomic-type orbitals. The STO-3G basis set is a minimal basis set (Collins et al. 1976).

Split valence basis sets, such as 3-21G and 6-31G, have two (or more) sizes of basis function for each valence orbital and allow orbitals to change size, but not to change shape (Binkley et al. 1980, Gordon et al. 1982, Pietro et al. 1982). *Triple split valence* basis sets, like 6-311G, use three sizes of contracted functions for each orbital-type (McLean and Chandler 1980, Krishnan et al. 1980).

Polarized basis sets allow orbitals to change their shape by adding orbitals with angular momentum beyond what is required for the description of each atom in the ground state. The 6-31G(d) basis set also known as 6-31G* which contains d functions added to the heavy atoms, is very popular for calculations involving up to medium-sized systems (Petersson et al. 1988, Petersson and Al-Laham 1991).

Diffuse functions are large-size versions of s- and p-type functions and allow orbitals to occupy a larger region of space. Basis sets with diffuse functions are important for systems where electrons are relatively far from the nucleus: molecules with lone pairs, anions, and other systems with significant negative charges, systems in their excited states, systems with low ionization potentials, and so on. The 6-31+G(d) is the 6-31G(d) basis set with diffuse functions added to heavy atoms (Petersson et al. 1988, Petersson and Al-Laham 1991).

2.5.5 Computational Details

The theoretical calculations of the structures and vibrational wavenumbers of all compounds investigated in the present work were performed by using the Gaussian 98 program package (Frisch et al. 1998). The DFT calculations were carried out with Becke's 1988 exchange functional (Becke 1988) and the Perdew–Wang 91 gradient corrected correlation functional (abbreviated as BPW91) (Perdew and Wang 1992) and Becke's three parameter hybrid method using the Lee–Yang–Parr correlation functional (abbreviated as B3LYP) (Becke 1993). For comparison purposes, *ab initio* calculations performed at the HF level of theory were also performed. The 6-31G*, 6-31 + G* and 6-311 + G* Pople basis sets were used for the geometry optimization and normal modes calculations at all theoretical levels.

Theoretical calculations of various Ag-molecule model compounds were done at the BPW91 theoretical level having in view previous results (Legge et al. 2001) that claimed the inferiority of the B3LYP functional as compared to the BPW methods in calculating silver-containing molecules. In this case "purposely tailored" basis sets have been employed, namely 6-31 + G* for the N, C, O, and H atoms and LanL2DZ for the Ag atom, which may have some advantages particularly as regards basis set completeness (Barone 1995).

At the optimized structures of the examined species no imaginary frequency modes were obtained proving that a local minimum on the potential energy surface was found.

References

Ahmadi TS, Wang ZL, Green TC, Henglein A, El-Sayed MA (1996) Shape-controlled synthesis of colloidal platinum nanoparticles. Science 28:1924–1926

Albrecht MG, Creighton JA (1977) Anomalously intense Raman spectra of pyridine at a silver electrode. J Am Chem Soc 99:5215–5217

Astilean S, Baia M, Maniu D, Pinzaru S, Iliescu T (2004) Fabrication of ordered noble-metal nanostructures via nanosphere lithography and their investigation as effective substrates for surface-enhanced Raman spectroscopy. Book of Abstracts of the International Bunsen Discussion Meeting Raman and IR spectroscopy in biology and medicine, Jena, 84

Astilean S, Baia M, Baia L, Farcau C, Toderas F (2005) Noble-metal films deposited on polystyrene colloidal crystal as effective substrate for surface-enhanced Raman spectroscopy. Book of Abstracts of the Surface Plasmon Photonics Confererence, Graz, 124

Astilean S, Baia M, Baia L, Farcau C, Maniu D (2006) Tunable surface-enhanced Raman scattering (SERS) from noble metal films deposited on polystyrene colloidal crystal and nanoparticle arrays fabricated by nanosphere litography. European Optical Society Topical Meeting on Molecular Plasmonic Devices, Engelberg, 74–76

Baia M, Baia L, Astilean S (2005) Gold nanostructured films deposited on polystyrene colloidal crystal templates for surface-enhanced Raman spectroscopy. Chem Phys Lett 404:3–8

Baia L, Baia M, Popp J, Astilean S (2006a) Gold films deposited over regular arrays of polystyrene nanospheres as highly effective SERS substrates from visible to NIR. J Phys Chem B 110:23982–23986

Baia M, Baia L, Popp J, Astilean S (2006b) Surface-enhanced Raman scattering efficiency of truncated tetrahedral Ag nanoparticle arrays mediated by electromagnetic couplings. Appl Phys Lett, doi: 10.1063/1.2193778

Baia M, Baia L, Popp J, Astilean S (2006c) Ordered metallic nanostructures obtained by nanosphere lithography as tunable SERS-active substrates. Book of Abstract of the 2nd International Conference Advanced Spectroscopies on Biomedical and Nanostructured Systems, Cluj-Napoca, 123

Baia M, Toderas F, Baia L, Popp J and Astilean S (2006d) Probing the enhancement mechanisms of SERS with p-aminothiophenol molecules adsorbed on self-assembled gold colloidal nanoparticles. Chem Phys Lett 422:127–132

Baker GA, Moore DS (2005) Progress in plasmonic engineering of surface-enhanced Raman-scattering substrates toward ultra-trace analysis. Anal Bioanal Chem 382:1751–1770

Barnes AJ (1977) Fourier Transform Spectroscopy. In Barnes AJ and Orville-Thomas WJ (eds) Vibrational spectroscopy-modern trends, Elsevier, Amsterdam

Barone V (1995) Structure, thermochemistry, and magnetic properties of binary copper carbonyls by a density-functional approach. J Phys Chem 99:11659–11666

Becke AD (1988) Density-functional exchange-energy approximation with correct asymptotic behavior. Phys Rev A 38:3098–3100

Becke AD (1993) A new mixing of Hartree-Fock and local density-functional theories. J Chem Phys 98:1372–1377

Bell RJ (1972) Introductory Fourier transform spectroscopy. Academic Press, New York

Binkley JS, Pople JA, Hehre WJ (1980) Self-consistent molecular orbital methods. 21. Small split-valence basis sets for first-row elements. J Am Chem Soc 102:939–947

Bjerneld EJ, Murty KVGK, Prikulis J, Käl M (2002) Laser-induced growth of Ag nanoparticles from aqueous solutions. Chem Phys Chem 3: 116–119

Brandt SE, Cotton TM (1993) Investigations of surfaces and interfaces. In: Rossiter BW, Baetzold RC (eds) Physical methods of chemistry series-Part B, vol. IXB, 2nd edn. Wiley & Sons, New York

Bunker PR (1979) Molecular symmetry and spectroscopy. Academic Press Inc., New York

Califano S (1976) Vibrational states. John Wiley & Sons, New York, London

Campion A, Ivanecky JE, Child CM, Foster M (1995) On the mechanism of chemical enhancement in surface-enhanced Raman scattering. J Am Chem Soc 117:11807–11808

Chalmers JM, Griffiths PR (2002) Handbook of vibrational spectroscopy, Wiley & Sons, Chichester, Baffins Lane

Collins JB, Schleyer PvR, Binkley JS, Pople JA (1976) Self-consistent molecular orbital methods. XVII. Geometries and binding energies of second-row molecules. A comparison of three basis sets. J Chem Phys 64:5142–5151

Cotton TM, Kim JH, Chumanov GD (1991) Application of surface-enhanced Raman spectroscopy to biological systems. J Raman Spectrosc 22:729–742

Creighton JA (1988) The selection rules for surface-enhanced Raman. In: Clark RH and Hester R (eds) Spectroscopy of Surface, Wiley & Sons, New York

Demtröder W (1981) Laser spectroscopy-basic concepts and instrumentation, Springer Verlag Berlin, Heidelberg

Dewar MJS, Reynolds CH (1986) An improved set of MNDO parameters for sulfur. J Comp Chem 2:140–143

Dieringer JA, McFarland AD, Shah NC, Stuart DA, Whitney AV, Yonzon CR. Young MA, Zhang X, Van Duyne RP (2006) Surface enhanced Raman spectroscopy: new materials, concepts, characterization tools, and applications. Faraday Discuss 132:9–26

Efrima SE (1986) Surface-enhanced Raman Scattering (SERS). In: Bockris JO'M, White R and Conway BE (eds) Modern aspects in electrochemistry, vol. 16, Plenum, New York

Emory SR, Nie S (1997) Near-field surface-enhanced Raman spectroscopy on single silver nanoparticles. Anal Chem 69:2631–2635

Felidj N, Aubard J, Levi G, Krenn JR, Salerno M, Schider G, Lamprecht B, Leitner A, Aussenegg FR (2002) Controlling the optical response of regular arrays of gold particles for surface-enhanced Raman scattering. Phys Rev B 65:075419-1–075419-9

Felidj N, Truong SL, Aubard J, Levi G, Krenn JR, Hohenau A, Leitner A, Aussenegg FR (2004) Gold particle interaction in regular arrays probed by surface enhanced Raman scattering. J Chem Phys 120: 7141–7146

Fleischmann M, Hendra PJ, McQuillann AJ (1974) Raman spectra of pyridine adsorbed at a silver electrode. Chem Phys Lett 26:163–166

Foresman B (1996) Ab initio techniques in chemistry: interpretation and visualization. In: Swift ML and Zielinski TJ (eds) What every chemist should know about computing. ACS Books, Washington DC

Frisch MJ, Trucks GW, Schlegel HB, Scuseria GE, Robb MA, Cheeseman JR, Zakrzewski VG, Montgomery JA Jr, Stratmann JE, Burant JC, Dapprich S, Millam JM, Daniels AD, Kudin KN, Strain MC, Farkas O, Tomasi J, Barone V, Cossi M, Cammi R, Mennucci B, Pomelli C, Adamo C, Clifford S, Ochterski J, Petersson GA, Ayala PY, Cui Q, Morokuma K, Malick DK, Rabuck AD, Raghavachari K, Foresman JB, Ciolowski JB, Ortiz JV, Stefanov BB, Liu G, Liashenko A, Piskorz P, Komaromi I, Gomperts R, Martin RL, Fox DJ, Keith T, Al-Laham MA, Peng CY, Nanayakkara A, Gonzales C, Challacombe M, Gill PMW, Johnson B, Chen W, Wong MW, Andres JL, Head-Gordon M, Repogle ES, Pople JA (1998) Gaussian 98. Revision A7. Gaussian Inc, Pittsburgh PA

Gao X, Davies JP, Weaver MJ (1990) A test of surface selection rules for surface-enhanced Raman scattering: the orientation of adsorbed benzene and monosubstituted benzenes on gold. J Phys Chem 94:6858–6865

Gesten J, Nitzan A (1980) Electromagnetic theory of enhanced Raman scattering by molecules adsorbed on rough surfaces. J Chem Phys 70:3023–3037

Gesten J, Nitzan A (1985) Photophysics and photochemistry near surfaces and small particles. Surf Sci 158:165–189

Gordon MS, Binkley JS, Pople JA, Pietro WJ, Hehre WJ (1982) Self-consistent molecular-orbital methods. 22. Small split-valence basis sets for second-row elements. J Am Chem Soc 104:2797–2803

Grand J, Kostcheev S, Bijeon JL ,Lamy de la Chapelle M, Adam PM, Rumyantseva A, Lerondel G, Royer P (2003) Optimization of SERS-active substrates for near-field Raman spectroscopy. Synthetic Metals 139:621–624

Grand J, Lamy de la Chapelle M, Bijeon JL, Adam PM, Vial A, Royer P (2005) Role of localized surface plasmons in surface-enhanced Raman scattering of shape-controlled metallic particles in regular arrays. Phys Rev B 72:033407-1–033407-4

Gunnarsson L, Bjerneld EJ, Xu H, Petronis S, Kasemo B, Käll M (2001) Interparticle coupling effects in nanofabricated substrates for surface-enhanced Raman scattering. Appl Phys Lett, 78:802–804

Guzonas DA, Irish DE, Atkinson GF (1990) Evidence for a photon-driven charge-transfer enhancement in the surface-enhanced Raman scattering of 1,4-diazabicyclo[2.2.2]octane at a silver electrode. Langmuir 6:1102–1112

Haynes CL, Van Duyne RP (2002) Plasmon scanned surface-enhanced Raman scattering excitation profiles. Mat Res Soc Symp Proc 728:S10.7.1–S10.7.6

Haynes CL, Van Duyne RP (2003) Plasmon-Sampled Surface-Enhanced Raman Excitation Spectroscopy. J Phys Chem B 107:7426–7433

Haynes CL, McFarland AD, Van Duyne RP (2005) Surface enhanced Raman spectroscopy. Anal Chem 340:339–346

Hehre WJ, Radom L, Schleyer PvR,. Pople JA (1986) Ab initio molecular orbital theory. Wiley, New York

Hildebrandt P, Stockburger M (1984) Surface-enhanced Raman spectroscopy of rhodamine 6G adsorbed on colloidal silver. J Phys Chem 88:5935–5944

Hildebrandt P, Stockburger M (1986) Surface enhanced resonance Raman study on fluorescein dyes. J Raman Spectrosc 17:55–58

Hirschfeld T (1976) Fellgett's Advantage in UV-VIS multiplex spectroscopy. Appl Spectrosc 30:68–69

Hirschfeld T, Chase B (1986) FT-Raman Spectroscopy: development and justification. Appl Spectrosc 40:133–137

Hirschfeld T, Schildkraut ER (1974) Fourier transform Raman spectroscopy. In: Lapp M and Penney C (eds) Laser Raman gas diagnostics, Plenum, New York

Hohenberg P, Kohn W (1964) Inhomogeneous electron gas. Phys Rev 136 B864–B871

Hollas JM (1992) Modern Spectroscopy, 2nd ed. Wiley & Sons, Baffins Lane

Hu X, Wang T, Wang L, Dong S (2007) Surface-enhanced Raman scattering of 4-amino-thiophenol self-assembled monolayers in sandwich structure with nanoparticle shape dependence: Off-surface plasmon resonance condition. J Phys Chem C 111:6962–6969

Inoue M, Ohtaka K (1989) Enhanced Raman scattering by metal spheres. I. Cluster effects. J Phys Soc Jpn 52:3853–3864

Jeanmaire DL, Van Duyne RP (1977) Surface Raman spectroelectrochemistry. Part I. Heterocyclic, aromatic, and aliphatic amines adsorbed on the anodized silver electrode. J. Electroanal Chem 84:1–20

Kim K, Lee HS, Kim NH (2007) Silver-particle-based surface-enhanced resonance Raman scattering spectroscopy for biomolecular sensing and recognition. Anal Bioanal Chem DOI 10.1007/s00216-007–1182-6

Kitaigorodski AI (1973) Molecular crystals and molecules. Academic Press, New York

Kneipp K, Kneipp H, Itzkan I, Dasari RR, Feld MS (2002) Surface-enhanced Raman scattering and biophysics. J Phys Condens Matter 14:597–624

Krishnan R, Binkley JS, Seeger R, Pople JA (1980) Self-consistent molecular orbital methods. XX. A basis set for correlated wave functions. J Chem Phys 72:650–654

Lee C, Yang W, Parr RG (1988) Development of the Colle-Salvetti correlation-energy formula into a functional of the electron density. Phys Rev B 37:785–789

Lee PC, Meisel D (1982) Absorption and surface-enhanced Raman of dyes on silver and gold sols. J Phys Chem 86:3391–3395

Legge FS, Nyberg GL, Peel JB (2001) DFT calculations for Cu-, Ag-, and Au-containing molecules. J Phys Chem A 105:7905–7916

Li X, Xu W, Zhang J, Jia H, Yang B, Zhao B, Li B, Ozaki Y (2004) Self-assembled metal colloid films: Two approaches for preparing new SERS active substrates. Langmuir 20:1298–1304

Li Y, Zhou J, Zhang K, Sunb C (2007) Gold nanoparticle multilayer films based on surfactant films as a template: Preparation, characterization, and application. J Chem Phys 126:094706-1–094706-7

McFarland AD, Young MA, Dieringer JA, Van Duyne RP (2005) Wavelength-scanned surface-enhanced Raman excitation spectroscopy. J Phys Chem B 109:11279–11285

McLean AD, Chandler GS (1980) Contracted Gaussian basis sets for molecular calculations. I. Second row atoms, $Z = 11–18$. J Chem Phys 72:5639–5648

Morse PM (1929) Diatomic molecules according to the wave mechanics. II. vibrational levels. Phys Rev 34:57–64

Moskovits M (1982) Surface selection rules. J Chem Phys 77:4408–4416

Moskovits M (2005) Surface-enhanced Raman spectroscopy: a brief retrospective. J Raman Spectrosc 36 485–496

Moskovits M, Suh JS (1984) Surface selection rules for surface-enhanced Raman spectroscopy: calculations and application to the surface-enhanced Raman spectrum of phthalazine on silver. J Phys Chem 88:5526–5530

Murphy T, Schmidt H, Kronfeldt HD (1999) Use of sol-gel techniques in the development of surface-enhanced Raman scattering (SERS) substrates suitable for in situ detection of chemicals in sea-water. Appl Phys B 69:147–150

Nickel U, Zu Castell A, Poppl K, Schneider S.A (2000) Silver colloid produced by reduction with hydrazine as support for highly sensitive surface-enhanced Raman spectroscopy. Langmuir 16:9087–9091

Niquist RA (2001) Interpreting infrared, Raman, and NMR spectra. Academic Press, San Diego

Orendorff CJ, Gole A, Sau TK, Murphy CJ (2005) Surface-enhanced Raman spectroscopy of self-assembled monolayers: Sandwich architecture and nanoparticle shape dependence. Anal Chem 77:3261–3266

Otto A (1991) Surface enhanced Raman scattering of adsorbates. J Raman Spectrosc 22:743–752

Otto A, Mrozek I, Grabhorn H, Akermann W (1992) Surface-enhanced Raman scattering. J Phys Condens Matter 4: 1143–1212

Oldenberg SJ, Westcott SL, Averitt RD, Halas NJ (1999) Surface enhanced Raman scattering in the near infrared using metal nanoshell substrates. J Chem Phys 111: 4729–4735

Perdew JP, Wang Y (1992) Accurate and simple analytic representation of the electron-gas correlation energy. Phys Rev B 45:13244–13249

Pertsin AJ, Kitaigorodski AI (1987) The atom-atom potential method. Springer, Berlin

Petersson GA, Bennett A, Tensfeldt TG, Al-Laham MA, Shirley WA, Mantzaris J (1988) A complete basis set model chemistry. I. The total energies of closed-shell atoms and hydrides of the first-row elements. J Chem Phys 89:2193–2218

Petersson GA, Al-Laham MA (1991) A complete basis set model chemistry. II. Open-shell systems and the total energies of the first-row atoms. J Chem Phys 94:6081–6090

Petry R, Schmitt M, Popp J (2003) Raman spectroscopy—A prospective tool in the life sciences. Chem Phys Chem 4:14–30

Pietro WJ, Francl MM, Hehre WJ, Defrees DJ, Pople JA, Binkley JS (1982) Self-consistent molecular orbital methods. 24. Supplemented small split-valence basis sets for second-row elements. J Am Chem Soc 104:5039–5048

Sánchez-Gil JA, García-Ramos JV (1998) Calculations of the direct electromagnetic enhancement in surface enhanced Raman scattering on random self-affine fractal metal surfaces. J Chem Phys 108: 317–325

Schmidt JP, Cross SE, Buratto SK (2004) Surface-enhanced Raman scattering from ordered Ag nanocluster arrays. J Chem Phys 121:10657–10659

Schrader B (1995) General Survey of Vibrational Spectroscopy. In: Schrader B (ed.) Infrared and Raman Spectroscopy, Methods and Applications, VCH, Weinheim

Sequaris JM, Koglin E (1985) Subnanogram colloid surface-enhanced Raman spectroscopy (SERS) of methylated guanine on silica gel plates. Anal Chem 321:758–759

Slater, JC (1974) Quantum theory of molecules and solids, vol. 4: The self-consistent field for molecules and solids. McGraw-Hill, New York

Stewart JJP (1989) Optimization of parameters for semiempirical methods II. Applications. J Comp Chem 10:221–264

Szabo A, Ostlund NS (1982) Modern quantum chemistry. McGraw-Hill, New York

Rivas L, Sanchez-Cortes S, Garcia-Ramos JV, Morcillo G (2001) Growth of silver colloidal particles obtained by citrate reduction to increase the Raman enhancement factor. Langmuir 17:574–577

Rosi NL, Mirkin CA (2005) Nanostructures in biodiagnostics. Chem Rev 105:1547–1562

Turkevich J, Stevenson PC, Hillier (1951) A study of the nucleation and growth processes in the synthesis of colloidal gold. Discuss Faraday Soc 11:55–75

Vlckova B, Gu XJ, Tsai DP, Moskovits M (1996) A microscopic surface-enhanced Raman study of a single adsorbate-covered colloidal silver aggregate. J Phys Chem 100:3169–3174

Vo-Dinh T (1988) Surface-enhanced Raman spectroscopy using metallic nanostructures. Trends in Anal Chem 17:557–582

Vo-Dinh T, Stokes DL, Griffin GD, Volkan M, Kim UJ, Simon MI (1999) Surface-enhanced Raman scattering (SERS) method and instrumentation for genomics and biomedical analysis. J Raman Spectrosc 30:785–793

Vosko SH, Wilk L, Nusair M (1980) Accurate spin-dependent electron liquid correlation energies for local spin density calculations: a critical analysis. Canadian J Phys 58:1200–1211

Xu H, Aizpurua J, Käll M, Apell P (2000) Electromagnetic contributions to single-molecule sensitivity in surface-enhanced Raman scattering. Phys Rev E 62:4318–4324

Zeiri L, Bronk BV, Shabtai Y, Czege J, Efrima S (2002) Silver metal induced surface enhanced Raman of bacteria. Colloids and Surfaces A: Physicochemical and Engineering Aspects 208:357–362

Zhou Q, Zhao G, Chao Y, Li Y, Wu Y, Zheng J (2007) Charge-transfer induced surface-enhanced Raman scattering in silver nanoparticle assemblies. J Phys Chem C 111 :1951–1954

Zhou Q, Li X, Fan Q, Zhang X, Zheng J (2006) Charge transfer between metal nanoparticles interconnected with a functionalized molecule probed by surface-enhanced Raman spectroscopy. Angew Chem Int Ed 45:3970–3973

Wang W, Gu B (2005) New surface-enhanced Raman spectroscopy substrates via self-assembly of silver nanoparticles for perchlorate detection in water. Appl Spectrosc 59:1509–1515

Wei H, Li J, Wang Y, Wang E (2007) Silver nanoparticles coated with adenine: preparation, self-assembly and application in surface-enhanced Raman scattering. Nanotechnology 18:175610-1–175610-5

Wilson EB, Decius JC, Cross PC (1955) Molecular vibrations, the theory of infrared and Raman vibrational spectra. McGraw-Hill, New York

Weitz DA, Moskovits M, Creighton JA (1986) Surface-Enhanced Raman Scattering with Emphasis on the Liquid-Solid Interface. In: Hall RB and Ellis AB (eds), Chemistry and structure at interfaces, new laser and optical techniques, VCH, Florida

3 Tranquilizers and Sedatives

3.1 Phenothiazine Derivatives

Chemotherapeutic agents are usually designated and used according to their most predominant pharmacological activity. There are very few drugs, however, with a single specific function. Several studies have demonstrated the potential role of the phenothiazine and its derivatives as anti-tumor (Motohashi 1991), anti-viral (Bohn et al. 1983, Candurra et al. 1996) and antiplasmid agents (Ford et al. 1989, Motohashi et al. 1992). All chemical compounds possessing moderate to powerful antimicrobial properties have been grouped together under the common term "non-antibiotics." Several groups of workers have repeatedly reported on the existence of moderate to powerful antimicrobial property in a variety of non-antibiotic compounds, particularly the phenothiazines (Dastidar et al. 2000, Shine and Mach 1965, Henry and Kasha 1967, Alkalis et al. 1975, Delay et al. 1952).

A new series of phenothiazine derivatives found to be important intermediates in the metabolism of phenothiazine drugs have been prepared (Tosa et al. 2001, Chetty et al. 1996) and the schematic structures of 10-isopentyl-10H-phenothiazine-5-oxide (10-I-10H-P-5-O) and 10-isopentyl-10H-phenothiazine-5,5-dioxide (10-I-10H-P-5,5-D) with the labeling of the atoms are illustrated in Fig. 3.1 (reprinted with permission from J. Phys. Chem. A 2003, 107, 1811–1816, copyright 2003 American Chemical Society and from Chem. Phys., 298, Bolboaca M, Iliescu T, Kiefer W, Infrared absorption, Raman, and SERS investigations in conjunctions with theoretical simulations on a phenothiazine derivative, 87–95, copyright 2004, with permission from Elsevier).

Raman spectra of phenothiazine and its radical cation were reported by Pan and Phillips (Pan and Phillips 1999), while Hester and Williams (Hester and Williams 1981) have reported the resonance Raman spectra of phenothiazine, 10-methylphenothiazine, and their radical cations.

In the next paragraphs a fairly detailed experimental and theoretical investigation of the newly prepared 10-I-10H-P-5-O and 10-I-10HP-5,5-D derivatives is pre-

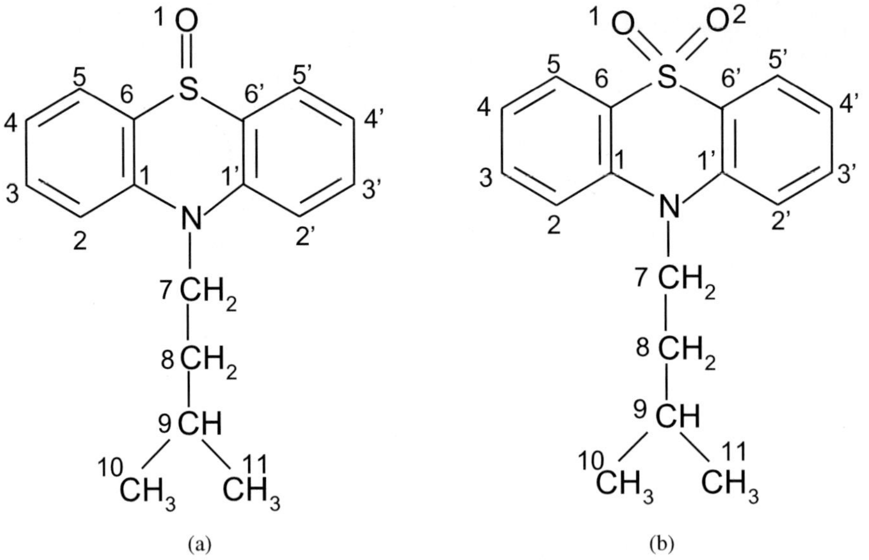

Fig. 3.1 Schematic structure of the 10-isopentyl-10H-phenothiazine-5-oxide (*a*) and 10-isopen-tyl-10H-phenothiazine-5,5-dioxide (*b*) compounds with the labeling of the atoms. Reprinted with permission from J. Phys. Chem. A 2003, 107, 1811–1816, copyright 2003 American Chemical Society (a) and from Chem. Phys., 298, Bolboaca M, Iliescu T, Kiefer W, Infrared absorption, Raman, and SERS investigations in conjunctions with theoretical simulations on a phenothiazine derivative, 87–95, copyright 2004, with permission from Elsevier (b)

sented. Firstly, the vibrational analysis of the most stable conformer of the title compounds is performed by means of infrared absorption and Raman spectroscopy in combination with theoretical (HF and DFT) simulations. Secondly, the SERS spectra at different pH values are analyzed in order to elucidate the adsorption behavior of the molecules on colloidal silver particles and to find out the pH influence.

3.1.1 Vibrational Analysis

Due to the flexibility of the isopentyl group, both phenothiazine derivatives 10-I-10H-P-5-O and 10-I-10H-P-5,5-D allow for several conformers. The optimized geometries of their six most probable conformers calculated at the BPW91/6-31G* level of theory are illustrated in Fig. 3.2.

Analytical harmonic vibrational modes have also been calculated to ensure that the optimized structures correspond to minima on the potential energy surface. The total energy of the most stable conformers, which were found to be the conformers 1, including zero point corrections, are −1186.6367 and −1261.8341 Hartree, respectively. The differences between the energy of the most stable conformer and the energy of the other relevant conformers, obtained at this theoretical

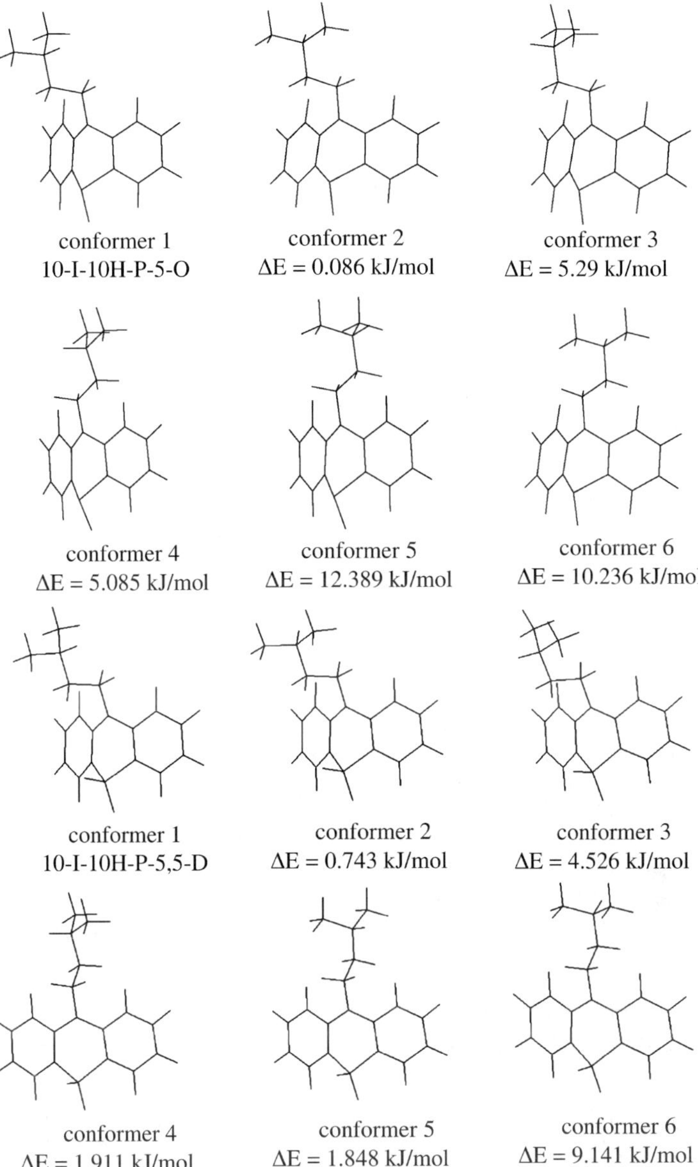

Fig. 3.2 Optimized geometries of the six most probable conformers of 10-I-10H-P-5-O and 10-I-10H-P-5,5-D. The differences between the energy of the most stable conformer and the energy of the other conformers are also indicated. Reprinted with permission from J. Phys. Chem. A 2003, 107, 1811–1816, copyright 2003 American Chemical Society (10-I-10H-P-5-O) and from Chem. Phys., 298, Bolboaca M, Iliescu T, Kiefer W, Infrared absorption, Raman, and SERS investigations in conjunctions with theoretical simulations on a phenothiazine derivative, 87–95, copyright 2004, with permission from Elsevier (10-I-10H-P-5,5-D).

level, are also indicated in Fig. 3.2. The experimental and theoretical investigations were further carried out for the conformer 1 of both phenothiazine derivatives, that will be further denoted as 10-I-10H-P-5-O and 10-I-10H-P-5,5-D, respectively.

According to the X-ray diffraction investigations (McDowell 1976) the phenothiazine molecule is folded about the N-S axis with the two planes containing the phenyl rings having a dihedral angle of 158.5°. It was reported (Pan and Phillips 1999) that the amount of folding increases for larger substituents on 10-substituted derivatives, chlorpromazine having a dihedral angle of 139.4°. Selected optimized structural parameters of both phenothiazine derivatives calculated by various methods are given in Table 3.1 (Bolboaca et al. 2003, Bolboaca et al. 2004) along with the available X-ray values of the ground state of the phenothiazine (McDowell 1976). As it can be observed, the theoretical dihedral angle between the two phenyl rings of both compounds has smaller values compared to the dihedral angle of the phenothiazine and agrees with previous findings (Pan and Phillips 1999). The calculated bond lengths and bond angles are in good agreement with the reported parameters (McDowell 1976), the B3LYP method giving the best results. At this level of calculation, the differences between the theoretical and experimental values of the structural parameters that involve the S and N atoms are mainly due to the substituent's presence.

FT-Raman and infrared spectra of the phenothiazine derivatives 10-I-10H-P-5-O and 10-I-10H-P-5,5-D in the range from 3400 to 400 cm^{-1} are presented in Fig. 3.3. The observed bands as well as the vibrational assignment performed with the help of the results obtained from theoretical simulations and the work of Pan and Phillips (Pan and Phillips 1999) are summarized in Table 3.2.

The neglect of anharmonicity effects and the incomplete incorporation of electron correlation in the ab initio theoretical treatment lead to harmonic vibrational wavenumbers larger than the fundamentals experimentally observed (Hehre et al. 1986). Having in view that Hartree–Fock (HF) calculations overestimate relatively uniform vibrational wavenumbers because of improper dissociation behavior, the predicted wavenumber values have to be scaled with scaling factors to adjust the observed experimental values (Scott and Radom 1996). Thus, the restricted HF (RHF) calculated vibrational wavenumbers presented in Table 3.2 have been uniformly scaled by 0.8953 according to the work of Scott and Radom (Scott and Radom 1996). Even after scaling, in comparison to the experiment, the RHF wavenumbers are overestimated in the high wavenumbers region, but are comparable to the experimental values in the low wavenumbers region.

In agreement with previous studies (Scott and Radom 1996, Wong 1996) the vibrational wavenumbers calculated using the B3LYP functional are also much larger than those calculated with the BPW91 method compared to the experimental values (see Table 3.2). Thus, according to the work of Rauhut and Pulay (Rauhut and Pulay 1995) a scaling factor of 0.963 has been uniformly applied to the B3LYP calculated wavenumbers from Table 3.2. The observed disagreement between the theory and experiment could be a consequence of the anharmonicity and of the general tendency of the quantum chemical methods to overestimate the

Table 3.1 Selected calculated bond lengths (Å) and angles (°) of the 10-I-10H-P-5-O and 10-I-10H-P-5,5-D derivatives compared to the experimental data of the phenothiazine (PhT)

	10-I-10H-P-5-O			10-I-10H-P-5,5-D			PhT
	Calc.[a]	Calc.[b]	Calc.[c]	Calc.[a]	Calc.[b]	Calc.[c]	Exp.[d]
Bond lengths (Å)							
$C\text{-}S_{average}$	1.821	1.821	1.809	1.784	1.784	1.774	1.77
$C\text{-}N_{average}$	1.419	1.419	1.417	1.409	1.409	1.406	1.406
$C_1\text{-}C_2$	1.412	1.412	1.404	1.415	1.415	1.408	1.385
$C_2\text{-}C_3$	1.402	1.402	1.395	1.398	1.398	1.391	1.39
$C_3\text{-}C_4$	1.402	1.402	1.395	1.405	1.310	1.398	1.367
$C_4\text{-}C_5$	1.401	1.401	1.395	1.397	1.397	1.390	1.367
$C_5\text{-}C_6$	1.396	1.396	1.389	1.401	1.401	1.395	1.391
$C_6\text{-}C_1$	1.413	1.413	1.405	1.417	1.417	1.409	1.397
$C_2\text{-}H_2$	1.091	1.091	1.084	1.091	1.091	1.084	0.98
$C_3\text{-}H_3$	1.093	1.093	1.086	1.093	1.093	1.086	1.05
$C_4\text{-}H_4$	1.092	1.092	1.085	1.092	1.092	1.085	0.98
$C_5\text{-}H_5$	1.093	1.093	1.086	1.092	1.092	1.085	0.93
$S\text{-}O_1$	1.522	1.522	1.512	1.486	1.486	1.427	–
$S\text{-}O_2$	–	–	–	1.488	1.488	1.474	–
$C_7\text{-}N$	1.471	1.471	1.466	1.477	1.477	1.473	–
Angles (°)							
Dihedral angle	138.478	138.478	137.657	145.814	145.788	145.253	153.30
$C_6\text{-}S\text{-}C_{6'}$	93.068	93.068	93.676	98.191	98.191	98.571	99.60
$C_1\text{-}N\text{-}C_{1'}$	117.207	117.206	116.922	120.639	120.639	120.475	121.50
$C_1\text{-}C_2\text{-}C_3$	120.484	120.484	120.417	120.716	120.717	120.658	119.8
$C_2\text{-}C_3\text{-}C_4$	121.091	121.092	121.084	121.280	121.280	121.297	120.5
$C_3\text{-}C_4\text{-}C_5$	119.165	119.165	119.192	119.015	119.015	119.016	119.4
$C_4\text{-}C_5\text{-}C_6$	119.538	119.538	119.502	119.720	119.719	119.662	119.7
$C_5\text{-}C_6\text{-}C_1$	122.307	122.307	122.268	122.267	122.267	122.299	119.2
$C_6\text{-}C_1\text{-}C_2$	117.377	117.377	117.501	116.958	116.958	117.025	119.5
$C_1\text{-}C_2\text{-}H_2$	120.267	120.267	120.302	120.139	120.139	120.126	118.5
$C_2\text{-}C_3\text{-}H_3$	118.874	118.873	118.915	118.409	118.786	118.823	115.8
$C_3\text{-}C_4\text{-}H_4$	120.554	120.554	120.520	120.604	120.604	120.580	117
$C_4\text{-}C_5\text{-}H_5$	121.874	121.874	121.842	121.729	121.730	121.683	122.8
$C_6\text{-}S\text{-}O_1$	110.020	110.020	109.540	110.035	110.035	109.955	–
$C_6\text{-}S\text{-}O_2$	–	–	–	108.564	108.564	108.662	–
$C_{6'}\text{-}S\text{-}O_1$	110.065	110.650	109.567	110.076	110.076	109.994	–
$C_{6'}\text{-}S\text{-}O_2$	–	–	–	108.428	108.429	108.526	–

Abbreviations: [a] Calculated with RHF/6-31G*, [b] Calculated with BPW91/6-31G*, [c] Calculated with B3LYP/6-31G*, d Ref. (McDowell 1976). Reprinted with permission from J. Phys. Chem. A 2003, 107, 1811–1816, copyright 2003 American Chemical Society (10-I-10H-P-5-O) and from Chem. Phys., 298, Bolboaca M, Iliescu T, Kiefer W, Infrared absorption, Raman, and SERS investigations in conjunctions with theoretical simulations on a phenothiazine derivative, 87–95, copyright 2004, with permission from Elsevier (10-I-10H-P-5,5-D)

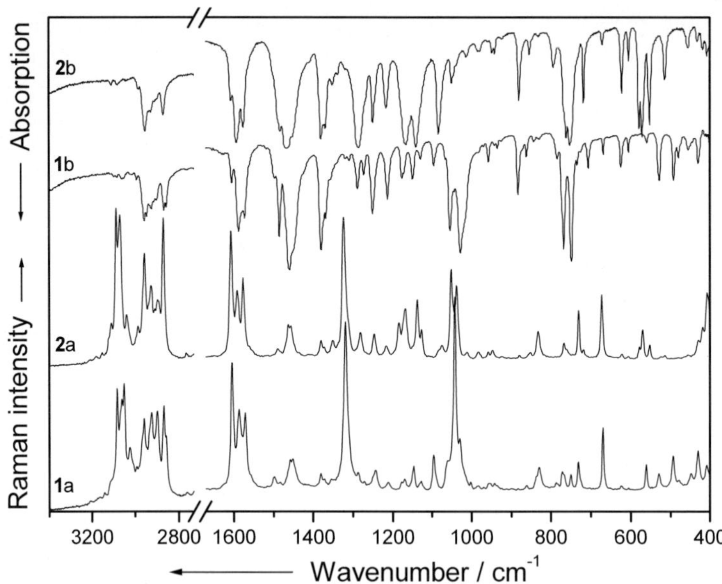

Fig. 3.3 FT-Raman (*a*) and infrared (*b*) spectra of 10-I-10H-P-5-O (*1*) and 10-I-10H-P-5,5-D (*2*) derivatives. Reprinted with permission from J. Phys. Chem. A 2003, 107, 1811–1816, copyright 2003 American Chemical Society (10-I-10H-P-5-O) and from Chem. Phys., 298, Bolboaca M, Iliescu T, Kiefer W, Infrared absorption, Raman, and SERS investigations in conjunctions with theoretical simulations on a phenothiazine derivative, 87–95, copyright 2004, with permission from Elsevier (10-I-10H-P-5,5-D)

force constants at the exact equilibrium geometry (Rauhut and Pulay 1995). However, as can be seen from Table 3.2, the theoretical results reproduce well the experimental data and allow the assignment of the vibrational modes.

By analyzing Fig. 3.3 and Table 3.2, one can remark that bands given by the CH stretching vibration of the phenyl ring and isopentyl group dominate the high wavenumber region (3200–2800 cm⁻¹) of the infrared and Raman spectra of both compounds. The stretching vibrations of the phenyl rings give rise to bands present in the range between 1610 and 1570 cm⁻¹ of all spectra. The strong Raman bands around 1320 cm⁻¹ and their corresponding infrared bands were also attributed to the CC stretching vibrations of the ring. The phenyl ring breathing vibration gives rise to the medium intense infrared and Raman bands at 1042 cm⁻¹ (calc. 1042 cm⁻¹) in the spectra of the 10-I-10H-P-5-O compound and at 1049 and 1051 cm⁻¹ (calc. 1043 cm⁻¹) in the infrared and Raman spectra of the 10-I-10H-P-5,5-D derivative. The bands that occur around 880 cm⁻¹ and in the spectral range between 675 and 600 cm⁻¹ of both spectra are due to the in-plane deformation vibrations of the phenyl rings, while the out-of-plane deformation vibrations appear at 705 (calc. 710 cm⁻¹), 530 (calc. 514 cm⁻¹), and 430 cm⁻¹ (calc. 440 cm⁻¹) in the infrared and Raman spectra of the 10-I-10H-P-5-O derivative and at 430 (calc. 441 cm⁻¹), 407 (calc. 395 cm⁻¹) and in the 580–570 cm⁻¹ spectral region of both spectra of the 10-I-10H-P-5,5-D compound. The medium intense Raman bands around 1248 cm⁻¹ and

Table 3.2 Assignment of the theoretical wavenumber values (cm^{-1}) to the experimental bands of the 10-I-10H-P-5-O and 10-I-10H-P-5,5-D derivatives

10-I-10H-P-5-O					10-I-10H-P-5,5-D					Vibrational assignment
Infrared	Raman	Calc.[a]	Calc.[b]	Calc.[c]	Infrared	Raman	Calc.[a]	Calc.[b]	Calc.[c]	
–	177 m	193	185	196	–	167 w	174	169	177	C_1NC_1', C_6SC_6' twist + SO def
–	199 m	203	202	208	–	187 m	194	197	203	CCC skel def
–	–	–	–	–	–	249 m	240	237	248	O_1SO_2 def
–	308 w	320	305	320	–	307 w	296	290	301	Ring chair def
–	340 m	338	333	345	–	336 m	335	334	346	C_1NC_1', C_6SC_6' twist
–	383 m	372	375	387	–	–	–	–	–	SO def + $C_{10,9,11}$ def
402 m	408 m	408	403	399	407 m	407 m	399	395	410	Out-of-plane Ph ring def + O_1SO_2 wag + $C_{10,9,11}$ def
–	–	–	–	–	418 m	418 sh	431	429	444	
430 m	430 m	438	440	436	432 m	430 sh	450	441	460	Out-of-plane Ph ring def
455 sh	448 sh	461	455	455	455 m	457 vw	460	455	471	$C_{7,8,9}$ def + CH def (CH_2, CH_3)
479 m	480 sh	497	490	471	–	–	–	–	–	$C_1NC_{1'}$, $C_6SC_{6'}$ wag
492 m	494 m	518	508	507	–	–	–	–	–	
–	–	–	–	–	515 m	514 vw	549	538	561	OSO bend
527 m	530 w	522	514	529	–	–	–	–	–	Out-of-plane Ph ring def
560 w	560 m	566	553	534	552 s	553 w	569	548	571	Ring chair def
–	–	–	–	–	571 s	569 m	572	555	578	Out-of-plane Ph ring def
–	–	–	–	–	580 s	578 sh	592	594	613	
604 w	605 vw	606	608	609	605 m	605 vw	603	607	627	In-plane Ph ring def
668 w	670 m	685	666	682	670 w	672 m	725	698	729	
705 m	–	733	710	693	718 ms	718 sh	736	721	749	Out-of-plane Ph ring def

Table 3.2 (Continued)

10-I-10H-P-5-O					10-I-10H-P-5,5-D					Vibrational assignment
734 sh	730 m	752	736	754	–	730 m	750	739	768	CH wag (ring) + CH def (CH_2)
747 s	749 w	768	749	769	751 s	761 sh	768	751	777	
767 s	771 w	776	765	778	760 s	766 w	772	762	787	
829 vw	830 m	804	823	843	832 w	831 m	801	822	841	$C_{10,9,11}$ stretch + CH twist (ring)
844 vw	835 sh	859	830	874	855 m	852 w	864	833	871	
881 m	882 vw	884	871	896	880 m	879 vw	878	874	899	$C_{3,4,5}$, $C_{3',4',5'}$ bend
941 sh	943 vw	936	939	948	943 m	947 vw	937	940	965	CH twist (ring) + CH def (CH_3)
959 vw	956 vw	963	949	982	950 w	958 w	960	947	987	
1003 sh	1002 vw	1003	1012	1042	1012 m	1011 w	1006	1009	1040	$C_{1,2,3}$ + $C_{3,4,5}$ bend
1027 s	1030 sh	1015	1031	1058	1040 sh	1038 m	1019	1026	1053	$C_{7,8}$ stretch + OSO stretch
1042 sh	1042 s	1021	1042	1067	1049 m	1051 m	1025	1043	1070	Ph ring breathing
1053 s	1059 sh	1041	1059	1082	1084 s	1075 w	1055	1051	1085	CH bend (ring) + $C_6SC_{6'}$ stretch + NC_7 stretch
1095 m	1095 m	1092	1087	1116	–	–	–	–	–	SO stretch
1128 w	1127 w	1114	1124	1154	–	1126 m	1112	1125	1156	CH def (CH_2, CH_3) + CH bend (ring) + $C_{10,9,8}$ stretch
1150 m	1142 m	1146	1145	1180	1141 s	1136 m	1134	1138	1170	
–	1170 w	1188	1170	1199	1167 s	1169 m	1166	1169	1199	
1214 m	1212 w	1211	1213	1248	1218 m	1217 w	1213	1217	1254	$C_1NC_{1'}$ as. stretch + CH bend (ring) + CH def (CH_2)
1251 m	1244 m	1249	1247	1283	1250 m	1248 m	1251	1245	1281	$C_1NC_{1'}$ s. stretch + CH rock (ring)
1273 m	1272 vw	1276	1272	1314	1287 s	1281 m	1285	1281	1322	CH def (CH_2) + CH rock (ring) +
1289 m	1288 sh	1281	1278	1318	–	–	–	–	–	NC_7 stretch + OSO stretch
1319 vw	1319 s	1325	1323	1350	1319 sh	1324 s	1306	1307	1345	CCC stretch (Ph ring)
–	–	–	–	–	1338 sh	1334 sh	1336	1333	1354	

Table 3.2 (Continued)

10-I-10H-P-5-O					10-I-10H-P-5,5-D					Vibrational assignment
1359 sh	1355 sh	1345	1349	1370	1352 sh	1351 m	1348	1352	1377	CH def (CH_2, CH_3) + CH rock (ring)
1381 s	1380 w	1396	1384	1431	1379 s	1380 w	1396	1383	1430	
1449 sh	1451 m	1458	1455	1499	1451 sh	1456 m	1460	1465	1507	$C_{6,1}$, $C_{6',1'}$ stretch
1460 s	1457 m	1467	1480	1522	1465 s	1462 m	1477	1483	1527	
1484 s	1484 vw	1477	1486	1529	1483 sh	1483 sh	1487	1486	1530	CH def (CH_2, CH_3)
1498 sh	1498 w	1496	1502	1543	–	1489 m	1494	1496	1539	
1572 m	1571 m	1586	1573	1622	1576 s	1576 m	1593	1580	1629	Ph ring stretch
1588 s	1587 m	1600	1591	1638	1592 s	1591 m	1606	1597	1645	
1605 m	1605 s	1615	1612	1658	1608 sh	1607 s	1618	1614	1661	
2868 m	2867 m	2882	2935	3004	2869 m	2869 s	2883	2963	3003	CH stretch (CH, CH_2, CH_3)
2925 m	2925 m	2951	3059	3051	2927 m	2927 m	2970	3071	3049	
2958 m	2958 m	3013	3069	3088	2855 m	2957 m	3016	3078	3091	
–	3025 m	3030	3086	3101	3040 vw	3039 m	3035	3095	3101	CH stretch (ring)
3051 vw	3050 s	3042	3121	3112	–	3070 s	3049	3124	3146	
3084 vw	3082 s	3048	3143	3190	3080 vw	3088 s	3052	3140	3193	

Abbreviations: [a]Calculated with RHF/6-31G*, [b]Calculated with BPW91/6-31G*, [c]Calculated with B3LYP/6-31G*, Ph = phenyl, w = weak, m = medium, s = strong, sh = shoulder, def = deformation, stretch = stretching, bend = bending, twist = twisting, wag = wagging, rock = rocking.
Reprinted with permission from J. Phys. Chem. A 2003, 107, 1811–1816, copyright 2003 American Chemical Society (10-I-10H-P-5-O) and from Chem. Phys., 298, Bolboaca M, Iliescu T, Kiefer W, Infrared absorption, Raman, and SERS investigations in conjunctions with theoretical simulations on a phenothiazine derivative, 87–95, copyright 2004, with permission from Elsevier (10-I-10H-P-5,5-D)

their corresponding infrared bands at $1250\,cm^{-1}$ were attributed to the symmetric CNC stretching vibration, while the weak Raman bands and the medium intense infrared bands around $1215\,cm^{-1}$ were assigned to the asymmetric CNC stretching vibration. The bands given by the CSC stretching vibration appear around 1055 and $1080\,cm^{-1}$ in the infrared and Raman spectra of the 10-I-10H-P-5-O and 10-I-10H-P-5,5-D derivatives, respectively. The ring chair deformation vibrations give rise to medium intense infrared and Raman bands around $555\,cm^{-1}$ and to the weak Raman band at $307\,cm^{-1}$. Other bands given by the CNC and CSC out-of-plane deformation vibrations appear around 490, 340, and $172\,cm^{-1}$, respectively.

The SO deformation and stretching vibrations give rise to the medium intense Raman bands at 383 (calc. $375\,cm^{-1}$) and $1095\,cm^{-1}$ (calc. $1087\,cm^{-1}$), respectively in the spectra of the 10-I-10H-P-5-O derivative (Bolboaca et al. 2003). The OSO stretching vibrations lead to the appearance of the medium intense Raman bands at 1281 (calc. $1281\,cm^{-1}$) and $1038\,cm^{-1}$ (calc. $1026\,cm^{-1}$) and their corresponding infrared bands at 1287 and $1040\,cm^{-1}$, respectively in the spectra of 10-I-10H-P-5,5-D compound (Bolboaca et al. 2004). The band observed at $383\,cm^{-1}$ (calc. $375\,cm^{-1}$) in the Raman spectrum of 10-I-10H-P-5-O compound was assigned to the SO deformation vibration, while the bands observed at 514 (calc. $538\,cm^{-1}$), 407 (calc. $395\,cm^{-1}$), and $307\,cm^{-1}$ (calc. $290\,cm^{-1}$) in the spectra of 10-I-10H-P-5,5-D compound were attributed to the deformation vibrations of the OSO group.

The other bands present in the infrared and Raman spectra of both phenothiazine derivatives (Bolboaca et al. 2003, Bolboaca et al. 2004) are mostly due to the vibrations of the isopentyl group.

3.1.2 Adsorption on the Silver Surface

The normal Raman spectra of polycrystalline 10-I-10H-P-5-O and 10-I-10H-P-5,5-D compounds are compared with the SERS spectra of the molecules adsorbed on silver colloid in Fig. 3.4.

SERS enhancements were detected only for molecules adsorbed on activated hydrosols, obtained by the co-adsorption of the chloride anions. The SERS activation of the colloids in the presence of chloride anions can be explained either in terms of an increased electromagnetic field or on the basis of a chemical enhancement mechanism (Hildebrandt et al. 1993). The assignment of the normal vibrational modes of the phenothiazine derivative to the SERS bands at different pH values is summarized in Table 3.3.

When a molecule binds to a metal surface, it can be either physisorbed or chemisorbed. In the case of physisorption (Moskovits 1985) the spectra of physisorbed and free molecules are similar. On the other hand, when the molecules are chemisorbed on the metal surface (Lombardi et al. 1986, Creighton 1983), the position and the relative intensities of the SERS bands are dramatically changed, due to the overlapping of the molecular and metal orbitals that leads to the formation of a new metal-molecule SERS complex. Both the electromagnetic mecha-

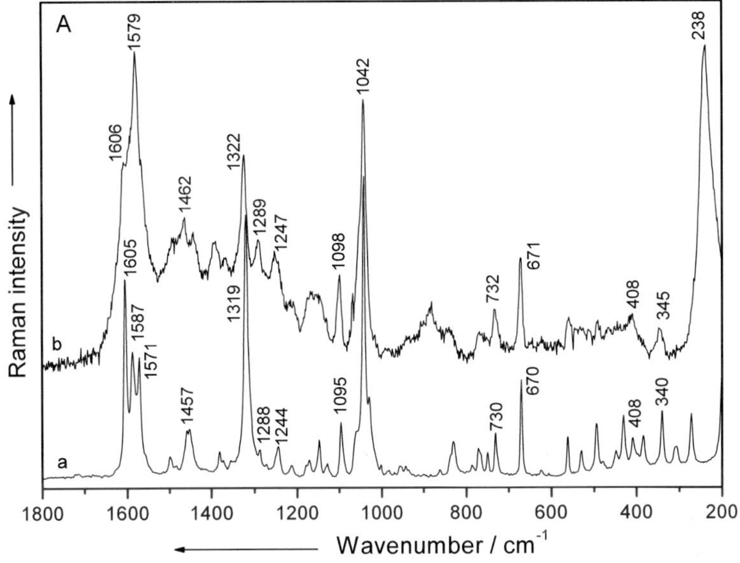
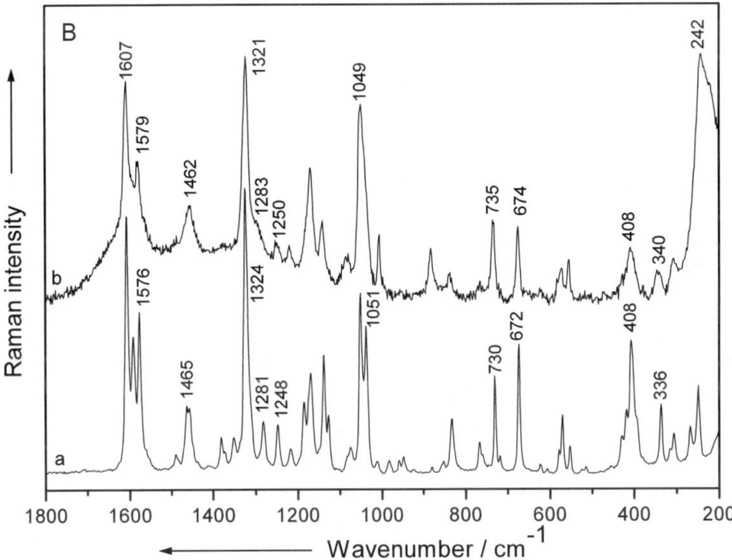

Fig. 3.4 FT-Raman (*a*) and SERS (*b*) spectra of 10-I-10H-P-5-O (**A**) and 10-I-10H-P-5,5-D (**B**) compounds. Reprinted from Chem. Phys., 298, Bolboaca M, Iliescu T, Kiefer W, Infrared absorption, Raman, and SERS investigations in conjunctions with theoretical simulations on a phenothiazine derivative, 87–95, copyright 2004, with permission from Elsevier (10-I-10H-P-5,5-D)

Table 3.3 Assignment of the normal vibrational modes of the 10-I-10H-P-5-O and 10-I-10H-P-5,5-D derivatives to the SERS bands

10-I-10H-P-5-O		10-I-10H-P-5,5-D		Vibrational assignment
Raman	SERS	Raman	SERS	
–	238 s	–	242 s	$AgCl^-$ stretch
–	–	307 w	306 m	Ring chair dcf
340 m	345 m	336 m	340 m	$C_1NC_{1'}$, $C_6SC_{6'}$ twist
408 m	408 m	408 m	408 m	Out-of-plane Ph ring def
494 m	493 m	–	–	$C_1NC_{1'}$, $C_6SC_{6'}$ wag
560 m	558 m	553 w	555 m	Ring chair def
–	–	569 m	571 m	Out-of-plane Ph ring def
670 m	671 m	672 m	674 m	In-plane Ph ring def
730 m	732 m	730 m	735 m	CH wag (ring) + CH def (CH_2)
771 w	768 m	–	–	
830 m	836 m	831 m	836 m	$C_{10,9,8}$ stretch + CH twist (ring)
884 vw	880 m	879 vw	881 m	$C_{3,4,5}$, $C_{3',4',5'}$ bend
–	–	1010 w	1005 m	$C_{1,2,3}$ + $C_{3,4,5}$ bend
1042 s	1042 s	1051 m	1049 s	Ph ring breathing
1059 sh	1067 sh	1074 w	1079 m	CH bend (ring) + $C_6SC_{6'}$ stretch + NC_7 stretch
1095 m	1098 m	–	–	SO stretch
1147 m	1153 m	1136 m	1140 m	CH bend (ring, CH_2, CH_3) + $C_{10,9,8}$ stretch
1170 w	1167 m	1169 m	1170 m	
1212 w	1213 m	1217 w	1219 w	$C_1NC_{1'}$ a. stretch + CH bend (ring) + CH def (CH_2)
1244 m	1247 m	1248 m	1250 w	$C_1NC_{1'}$ s. stretch + CH rock (ring)
1288 sh	1289 m	1281 m	1283 sh	CH def (CH_2) + CH rock (ring) + NC_7 stretch + OSO stretch
1319 s	1322 s	1324 s	1321 s	CCC stretch (Ph ring)
1380 w	1389 m	–	–	CH def (CH_2, CH_3) + CH rock (ring)
1451 m	1441 m	1456 m	1454 m	
1457 m	1462 m	1462 m	–	$C_{6,1}$, $C_{6',1'}$ stretch
1571 m	1579 s	1576 m	1579 m	Ph ring stretch
1587 m	1592 sh	1591 m	1596 sh	
1605 s	1606 sh	1607 s	1607 s	
2867 m	–	2869 s	2866 sh	CH stretch (CH, CH_2, CH_3)
2898 m	2871 s	2896 m	2896 sh	
2925 m	2935 s	2927 m	2938 s	
2958 m	2958 s	2957 m	2970 sh	
3050 s	–	3039 m	3036 sh	CH stretch (ring)
3061 sh	3067 s	3070 s	3071 s	
3082 s	3089 sh	3088 s	–	

Abbreviations: Ph = phenyl, w = weak, m = medium, s = strong, sh = shoulder, stretch = stretching, bend = bending, twist = twisting, wag = wagging, rock = rocking. Reprinted with permission from J. Phys. Chem. A 2003, 107, 1811–1816, copyright 2003 American Chemical Society (10-I-10H-P-5-O) and from Chem. Phys., 298, Bolboaca M, Iliescu T, Kiefer W, Infrared absorption, Raman, and SERS investigations in conjunctions with theoretical simulations on a phenothiazine derivative, 87–95, copyright 2004, with permission from Elsevier (10-I-10H-P-5,5-D)

nism and the charge transfer effect contribute to the overall SERS effect. In the case of physisorbed molecules, the electromagnetic mechanism is the main mechanism of the Raman enhancement, while in the case of chemisorption, the charge transfer effect is the dominant mechanism.

By looking at the geometry of the molecules, one can assume that they may bind to the silver surface either through the π orbitals of the phenyl rings or through the lone pair electrons of the nitrogen or oxygen atoms. For aromatic molecules, it is known (Gao and Weaver 1985) that the bands due to the ring vibrations are red shifted by more than $10\,cm^{-1}$ and their bandwidths increase substantially when the molecules adsorb on the metal surface via their π systems. Since the SERS spectra exhibit shifts that never exceed $5\,cm^{-1}$ compared to the normal Raman spectra and the bandwidths are hardly affected by the adsorption, it is likely that the molecules are adsorbed on the silver surface via the nonbonding electrons of either the nitrogen or oxygen atoms. When looking at the presence of the isopentyl substituent on the nitrogen atom, one can assume (Bolboaca et al. 2003, Bolboaca et al. 2004) that the molecule-surface interaction occurs through the lone pair electrons of the oxygen atom, the nitrogen-metal interaction being sterically hindered.

The adsorbate-metal interaction is further evidenced by the presence of some bands in the 250–$150\,cm^{-1}$ region (Fig. 3.5), which are mainly assigned to the Ag-adsorbate stretching vibrations (Sanchez-Cortes and Garcia-Ramos 1992, Chowdhury et al. 2000).

When the pair of the bands present in this spectral range passes from acidic to alkaline pH values, it shows intensity reversal. Thus, the intensity of the sharp band observed at pH$\,=\,$1 around $240\,cm^{-1}$, which is most probably due to the AgCl stretching vibration (Sanchez-Cortes and Garcia-Ramos 1992), is decreasing upon an increase of the pH values; meanwhile, an increase and a small shift to higher wavenumbers of the shoulder observed around $217\,cm^{-1}$ at pH$\,=\,$1 occurs, which corresponds to the AgO stretching vibration (Chowdhury et al. 2000). The high intensity of the band around $240\,cm^{-1}$ could be explained by the increase of the chloride anions concentration in an acidic environment caused by the addition of the acid for pH value adjustment. At pH$\,=\,$6, the AgCl$^-$ stretching band becomes broader and less intense, while the intensity of the AgO stretching band increases. At a still higher pH value (pH$\,=\,$14), as the concentration of the chloride ions is reduced in the medium and, consequently, at the silver surface, the AgCl$^-$ stretching vibration gives rise only to a shoulder around $240\,cm^{-1}$, while the intensity of the band at $221\,cm^{-1}$, which gives evidence of the metal-oxygen interaction, further increases. The appearance of the AgO stretching band in the SERS spectra at all pH values indicates the partial chemisorption of the molecules on the metal surface through the nonbonding electrons of the oxygen atom (Bolboaca et al. 2003, Bolboaca et al. 2004).

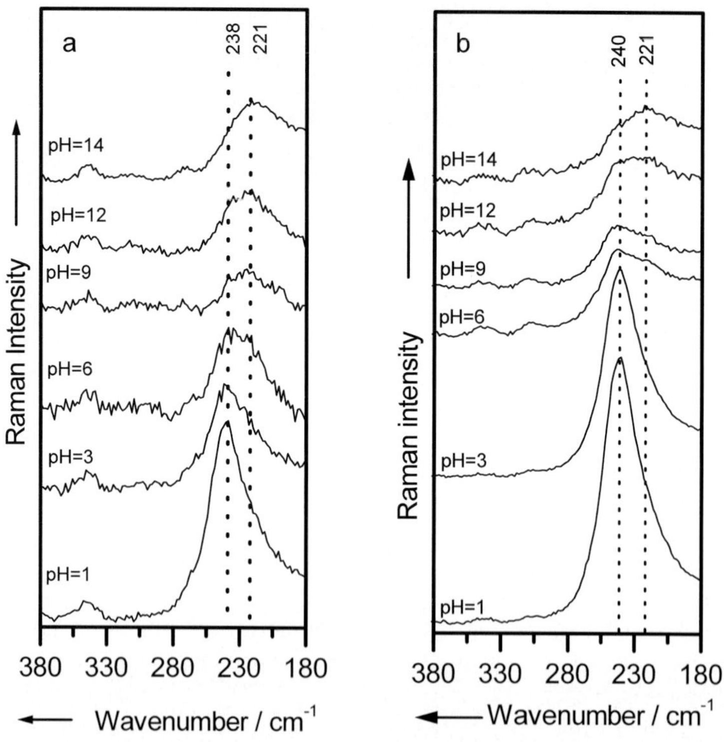

Fig. 3.5 pH dependence of the metal-adsorbate stretching mode from the SERS spectra of the (**a**) 10-I-10H-P-5-O and (**b**) 10-I-10H-P-5,5-D compounds Reprinted with permission from J. Phys. Chem. A 2003, 107, 1811–1816, copyright 2003 American Chemical Society (10-I-10H-P-5-O) and from Chem. Phys., 298, Bolboaca M, Iliescu T, Kiefer W, Infrared absorption, Raman, and SERS investigations in conjunctions with theoretical simulations on a phenothiazine derivative, 87–95, copyright 2004, with permission from Elsevier (10-I-10H-P-5,5-D)

The UV-vis absorption spectra of the pure colloid and mixture of the colloid and phenothazine derivatives before and after addition of NaCl were recorded and are presented in Fig. 3.6.

The absorption spectrum of the silver colloid shows a single absorption maximum at 407 nm, due to the small particle plasmon resonance. It is known (Muniz-Miranda 1999) that, when two metallic spheres approach each other, this band remains at the original single sphere wavelength, while another resonance develops at longer wavelengths; hence, a secondary peak occurs in the red/near-infrared (500–800 nm) spectral region. The appearance of such a new, broad band in the red/infrared region is generally attributed to the coagulation of silver particles in the sol in the presence of the adsorbed molecules (Fu and Zang 1992, Liang et al. 1993). Alternatively, such a band has been ascribed to a charge transfer band, due to the molecule-metal interaction (Sanchez-Cortes et al. 1995). One can see from Fig. 3.6 that, after the addition of both samples in silver hydrosol, the band at

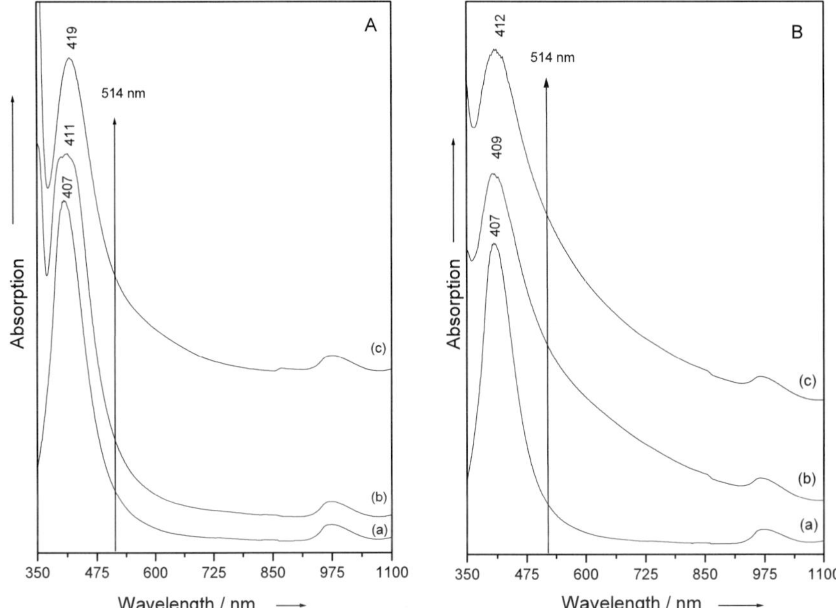

Fig. 3.6 Absorption spectra of salt-free silver colloidal dispersion (*a*), with 10^{-1} M 10-I-10H-P-5-O (**A**) and 10-I-10H-P-5,5-D (**B**) (*b*), with 10^{-1} M 10-I-10H-P-5-O (**A**) and 10-I-10H-P-5,5-D (**B**) and 10^{-2} M NaCl (*c*). Reprinted with permission from J. Phys. Chem. A 2003, 107, 1811–1816, copyright 2003 American Chemical Society (A) and from Chem. Phys., 298, Bolboaca M, Iliescu T, Kiefer W, Infrared absorption, Raman, and SERS investigations in conjunctions with theoretical simulations on a phenothiazine derivative, 87–95, copyright 2004, with permission from Elsevier (B)

407 nm becomes weaker and broader and is shifted to longer wavelengths by a few nanometers (below 4 nm). When NaCl is added to the colloid-sample mixture, the absorption peak is further shifted to longer wavelengths, and its intensity decreases, while no new band due to the secondary plasmon resonance is observed. This behavior indicates the significant contribution of the electromagnetic mechanism to the overall SERS enhancement.

Variations in the SERS spectra with the change of the pH values were usually attributed either to a change in orientation of adsorbates with respect to the metal surface (Takahashi et al. 1987) or to a change in the chemical nature of the adsorbates (Sun et al. 1985, Anderson and Evans 1988). By analyzing Fig. 3.7 one can see that no new bands occur in the SERS spectra at different pH values relative to the Raman spectrum, and therefore one can assume that the differences between the spectra originate from an orientational change of the molecule relative to the silver surface.

From the enhancement of relevant bands following the surface selection rules (Hallmark and Campion 1986, Moskovits and Suh 1984) one can predict the orientation of the adsorbed molecules with respect to the metal surface. According to

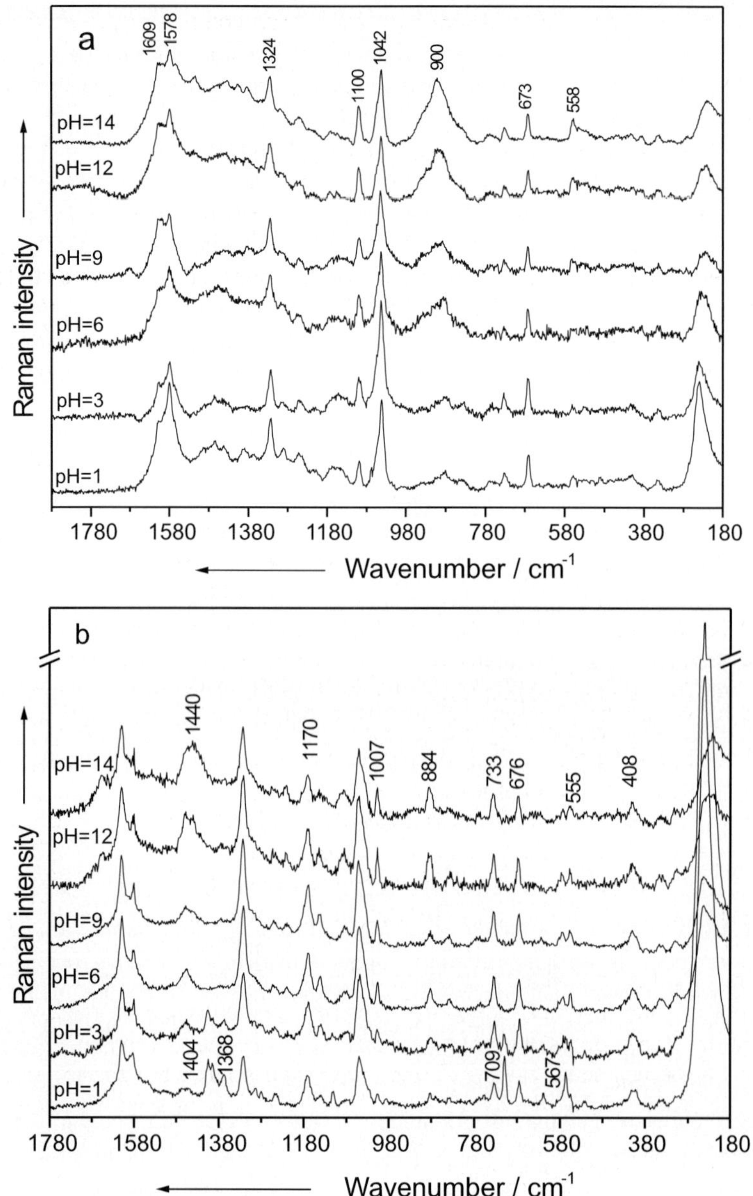

Fig. 3.7 SERS spectra of (**a**) 10-I-10H-P-5-O and (**b**) 10-I-10H-P-5,5-D compounds at different pH values as indicated. Reprinted with permission from J. Phys. Chem. A 2003, 107, 1811–1816, copyright 2003 American Chemical Society (a) and from Chem. Phys., 298, Bolboaca M, Iliescu T, Kiefer W, Infrared absorption, Raman, and SERS investigations in conjunctions with theoretical simulations on a phenothiazine derivative, 87–95, copyright 2004, with permission from Elsevier (b)

these rules, a vibrational mode with its normal mode component perpendicular to the surface will be more enhanced than a parallel one. Furthermore, the CH stretching vibrations were reported to the unambiguous probes for adsorbate orientations (Gao et al. 1990).

When looking at the geometry of the 10-I-10H-P-5-O derivative, it is very difficult to exactly predict the orientation of the molecule with respect to the metal surface. However, from Figs. 3.7a and 3.3a, one may notice that, at all pH values, the bands around 1600 and 1042 cm^{-1} assigned to the ring stretching and breathing modes, respectively, are enhanced, and small shifts of the bands, due to the ring stretching vibrations, compared to the corresponding Raman bands, are observed (Table 3.3).

The enhancement of the band around 670 cm^{-1} given by the in-plane deformation vibration of the aromatic ring can be seen at all pH values. At an alkaline pH, this band is shifted to higher wavenumbers by 3 cm^{-1} compared to the Raman spectrum. At all pH values, the bands at 408, 430, 530, and 703 cm^{-1}, which arise from the out-of-plane deformation vibrations of the phenyl rings, are very weakly enhanced. The behavior of the above-mentioned bands confirms the supposition that the molecule-metal interactions do not occur via the π orbitals of the aromatic rings. Moreover, the enhancement and the shift to higher wavenumbers by 3 cm^{-1} of the band at 1095 cm^{-1} due to the SO stretching vibration (Table 3.3) gives further evidence of the existence of the molecule-surface interaction through the lone pair electrons of the oxygen atom.

The enhancement of the bands due to the CH deformation and stretching vibrations of the isopentyl substituent present in all SERS spectra confirms the major contribution of the electromagnetic effect to the SERS enhancement, compared to the chemical one (Bolboaca et al. 2003).

When the SERS spectra at acidic and alkaline pHs are compared to the normal Raman spectrum (Figs. 3.7a and 3.3a), changes in the relative intensities on some bands can be noticed. Thus, at pH = 1 some bands due to the in-plane CH deformation vibration of the phenyl rings are enhanced. Furthermore, the enhancement of the bands at 1441 and 1462 cm^{-1}, given by the stretching vibrations of the C_6C_1 and $C_{6'}C_{1'}$ bonds, can be explained on the basis of surface selection rules (Sun et al. 1985, Anderson and Evans 1988). When the molecule interacts with the silver surface and adopts an upright orientation of the phenyl groups, these bonds are exactly perpendicular to the surface. For such a case, the surface selection rules predict particularly large enhancement, which is in agreement with our results. In the high wavenumber region (3200–3000 cm^{-1}), one can see the strong intensity of the CH stretching bands.

Therefore, one can assume that, at an acidic pH, the molecules are oriented at the metal surface in such a way that the benzene rings are preponderantly perpendicular with respect to the surface as indicated in Fig. 3.8a (Bolboaca et al. 2003).

According to the surface selection rules, (Moskovits and Suh 1984, Hallmark and Campion 1986) the out-of-plane vibrations are expected in the SERS spectrum only when the adsorbed molecules adopt a flat, or at least tilted, orientation on the silver surface. The increased intensity of the bands around 900 cm^{-1}, attributed to the out-of-plane CH deformation vibrations of the phenyl at an alkaline pH, can be

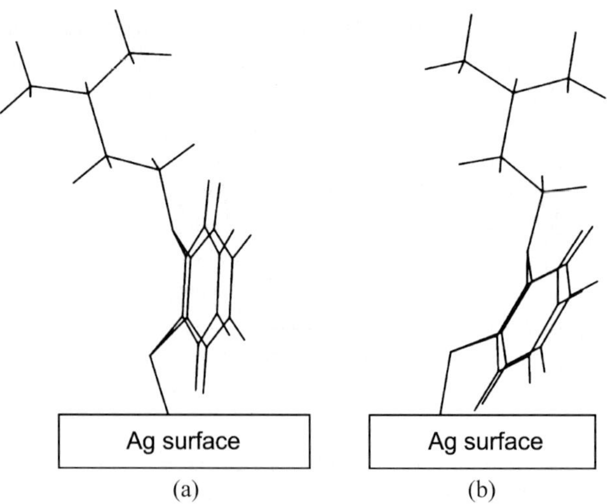

Fig. 3.8 Schematic model for the adsorption geometry of 10-I-10H-P-5-O on a colloidal silver surface at (**a**) acidic and (**b**) alkaline pHs. Reprinted with permission from J. Phys. Chem. A 2003, 107, 1811–1816. Copyright 2003 American Chemical Society

a consequence of the reorientation of the aromatic rings from upright to tilted. In contrast to the behavior revealed at an acidic pH, the bands assigned to the C_6C_1 and $C_{6'}C_{1'}$ stretching vibrations are only weakly enhanced at these values of the pH. Furthermore, the intensity of the CH stretching bands present in the high wavenumber region decreases. An enhancement of the band at $558\,cm^{-1}$, due to the ring chair deformation, is also observed. Therefore, one can suppose that, at an alkaline pH, a reorientation of the molecules occurs, with the phenyl rings obtaining a tilted orientation with respect to the metal surface, as suggested in Fig. 3.8b (Bolboaca et al. 2003).

By comparing the SERS spectra of the 10-I-10H-P-5,5-D derivative depicted in Fig. 3.7b, noticeable changes can be observed on passing from an acidic to an alkaline environment. Some bands are enhanced only in the SERS spectra at pH values below 6, while others appear only in the SERS spectra recorded in an alkaline environment, and therefore only the SERS spectra at pH values of 1 and 14 will be discussed.

Thus, in the SERS spectrum recorded at a pH value of 1, the band at $567\,cm^{-1}$, assigned to the out-of-plane deformation vibration of the phenyl ring, appears enhanced. The band at $709\,cm^{-1}$ attributed to the ring chair deformation vibration is also enhanced at this pH value. In the spectral range between 1404 and $1365\,cm^{-1}$ two bands due to the CH deformation vibrations of the CH_2 and CH_3 groups are enhanced in the SERS spectrum at pH = 1. At this pH value the bands due to the in-plane deformation vibration of the phenyl ring are not present or are only weakly enhanced. Having in view all these considerations, one can assume (Bolboaca et al. 2004) that in an acidic environment (pH<6) the molecule adopts a tilted orientation on the silver surface (see Fig. 3.9a).

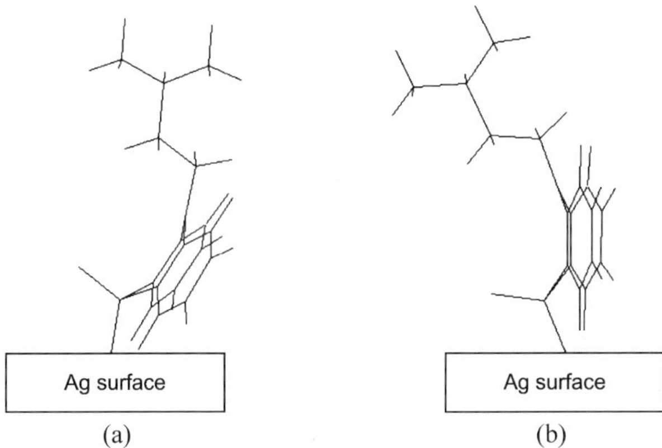

Fig. 3.9 Schematic model for the adsorption geometry of 10-I-10H-P-5,5-D on a colloidal silver surface at (**a**) pH < 6 and (**b**) pH > 6. Reprinted from Chem. Phys., 298, Bolboaca M, Iliescu T, Kiefer W, Infrared absorption, Raman, and SERS investigations in conjunctions with theoretical simulations on a phenothiazine derivative, 87–95, copyright 2004, with permission from Elsevier

By looking at the SERS spectrum at pH = 14 one can observe that in contrast to the spectra at an acidic pH, at this pH value the bands at 884 and 1007 cm^{-1} due to the in-plane deformation vibration of the phenyl ring are enhanced. The band at 1440 cm^{-1} attributed to the C$_6$C$_1$ and C$_6$OC$_{10}$ stretching vibrations also appears enhanced at this pH value. Furthermore, the bands at 1221 and 1254 cm^{-1} attributed to the CNC stretching vibration are also enhanced in the SERS spectrum recorded at pH = 14. At pH values above 6 the bands attributed to the CH stretching vibrations of the CH$_2$ and CH$_3$ groups are more enhanced compared to those obtained at an acidic pH.

By taking into account the behavior of these bands one can assume (Bolboaca et al. 2004) that in an alkaline environment (pH > 6) the 10-I-10H-P-5,5-D molecule is adsorbed on the metal surface in such a way that the phenyl rings have an upright orientation with respect to the surface (see Fig. 3.9b). At all pH values the adsorption of the molecule on the metal surface is maintained through the oxygen atom.

3.1.3 Conclusions

Throughout the study described in this section, experimental (infrared and Raman spectroscopy) and theoretical (HF and DFT calculations) investigations have been performed on the most stable conformers of 10-I-10H-P-5-O and 10-I-10H-P-5,5-D derivatives.

The SERS spectra of the samples in activated silver colloids were recorded and compared to the corresponding Raman spectra. The small shifts of the SERS bands ($v \leq 5\,cm^{-1}$) relative to the corresponding Raman bands and the presence of the Ag-molecule stretching band at all pH values led to the conclusion that both molecules are chemisorbed on the metal surface through the oxygen atom. The significant contribution of the electromagnetic mechanism to the overall SERS enhancement has been confirmed by the lack of a broad band in the long wavelengths region of the UV-vis spectra of the colloid with added adsorbate. The changes observed in the SERS spectra at different pH values were explained by considering the reorientation of the adsorbed molecule with respect to the silver surface.

3.2 Anthranil

Anthranil (2, 1-benzisoxazole) (reprinted with permission from J. Phys. Chem. B 2004, 108, 17491–17496. Copyright 2004 American Chemical Society) and its derivatives, whose infrared and Raman spectra have been reported some time ago (Millet et al. 1980), are well known due to their extensive applications in pharmacology and analytical chemistry (Armuth and Berenblum 1982, Manolov and Todorov 1974, Doppler et al. 1979). While the topical anthranil treatment was suggested for possible use in the bioassay of tumor promoters (Armuth and Berenblum 1982) and some anthranil derivatives have shown cytotoxic and mutagenic activity, other anthranil derivatives proved to possess protective antiulcer effects and sedative activity (Manolov and Todorov 1974). It was also shown (He et al. 1988) that anthranil is the key intermediate in the decomposition process of nitroaromatic explosives.

For understanding the action of potential drugs, such as anthranil, it is essential to find out if the structure of the adsorbed species is similar to that of the free molecule. In these investigations a silver surface may serve as an analogue for an artificial biological interface, and after elucidating the adsorption mechanism of a molecule, the study can be expanded to the adsorption on membranes or other interesting biological surfaces for medical or therapeutic treatments (Dryhurst 1977).

In the following paragraphs, SERS has been applied to the anthranil molecule in order to get insights about its adsorption behavior on the colloidal silver particles and to find out, from the enhancement of different Raman bands, the most probable orientation of the adsorbed species relative to the metal surface. DFT calculations have been also performed on different Ag-anthranil model complexes to provide valuable information concerning the interaction between the anthranil species and the silver surface.

3.2.1 Vibrational Analysis

The overall planar anthranil molecule consists of two planar cycles, a phenyl and an isoxazole ring, and belongs to the C_s point group symmetry. The schematic structure of this molecule with the labeling of the atoms is given in Fig. 3.10.

Mille and coworkers (Mille et al. 1980) have recorded and assigned the infrared and Raman spectra of the 2, 1-benzisoxazole molecule and compared them with those of the 1, 2-benzisoxazole compound. In this work, a more detailed assignment of the vibrational modes of the title compound based on the results of DFT calculations (Baia et al. 2004) is provided. The infrared and Raman spectra of the anthranil molecule in the spectral range between 450 and 3250 cm^{-1} are illustrated in Fig. 3.11.

The assignment of the observed bands, which is summarized in Table 3.4, was based on the visual inspection of normal mode displacement vectors by taking into account the vibrational bands positions and intensities. The proposed assignment is in agreement with the previously reported data (Mille et al. 1980).

By comparing the calculated vibrational wavenumbers with the experimental results (see Table 3.4) one can observe that, similar to the results of previous studies (Scott and Radom 1996, Wong 1996), the computed data using the B3LYP method are larger than those calculated with the BPW91 method. It is also noteworthy to mention that the theoretical simulations were performed for the gas phase, while the experimental data were obtained for a liquid sample.

However, as can be seen from Table 3.4, the predictions of the DFT methods for the vibrational frequencies are in consistently good agreement with the experimental values and allow for the complete assignment of the vibrational modes.

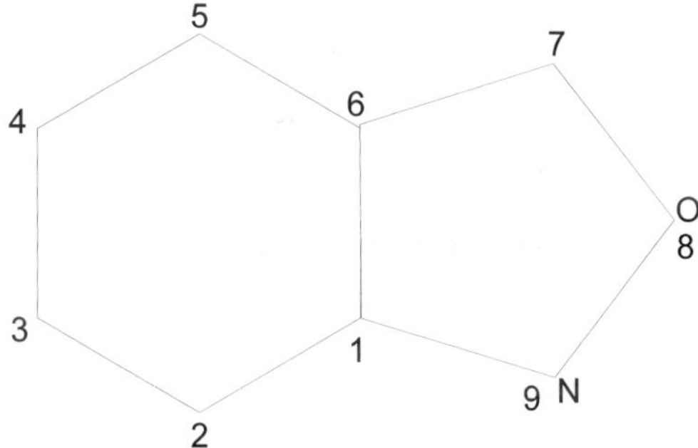

Fig. 3.10 Schematic structure of the anthranil molecule with the labeling of the atoms. Reprinted with permission from J. Phys. Chem. B 2004, 108, 17491–17496. Copyright 2004 American Chemical Society

Table 3.4 Assignment of the theoretical wavenumber values (cm^{-1}) to the experimental bands of the anthranil molecule

Experimental infrared	Raman	Theoretical Calc.[a]	Calc.[b]	Vibrational assignment
–	214 mw	206	212	Ring 1 + 2 out-of-plane def
–	264 w	244	253	
–	401 mw	394	407	$N_9C_{1,2} + C_{5,6,7}$ bend
–	435 w	427	442	Ring 1 out-of-plane def
–	536 m	529	544	$C_{1,2,3} + C_{4,5,6}$ bend
591 ms	592 sh	589	604	Ring 2 out-of-plane def
604 m	603 m	596	613	$N_9C_{1,6} + O_8C_{7,6}$ bend
734 m	741 sh	751	763	CH wag
752 s	751 ms	756	767	$C_{1,2,3} + C_{4,5,6} + C_7O_8N_9$ bend
809 m	812 vw	826	857	CH twist (ring 1)
868 m	869 mw	863	902	N_9O_8 stretch
906 m	905 m	887	918	$C_{1,2,3} + C_{3,4,5} + C_{4,5,6}$ bend
921 m	922 m	906	932	$C_7O_8N_9$ bend
956 w	957 vw	956	968	C_7H twist
981 m	980 m	993	1008	Ring 1 breathing
1113 s	1114 vw	1116	1155	C_7H twist
1141 m	1140 m	1133	1166	CH bend (ring 1)
1157 m	1156 m	1157	1190	
1238 m	1237 m	1234	1267	$C_{6,7}O_8$ stretch
1258 w	1258 m	1249	1292	CH rock (ring 1)
1352 m	1357 m	1348	1390	CH rock (ring 1) + C_1N_9 stretch
1381 s	1380 m	1365	1419	$C_1N_9 + C_{6,7}$ stretch
1410 m	1409 m	1398	1436	$C_{1,6} + C_{3,4}$ stretch
1456 m	1454 vs	1465	1503	$C_{2,3} + C_{4,5}$ stretch
1513 sh	1513 sh	1518	1563	$C_{1,2,3}$ s. stretch
1520 m	1518 m	–	–	
1555 m	1555 m	1555	1600	$C_{4,5,6}$ s. stretch
1641 s	1640 m	1637	1689	$C_{1,2,3} + C_{4,5,6} + C_{5,6,7}$ as. stretch
3013 m	3012 m	3123	3188	CH stretch (ring 1)
3068 m	3071 s	3133	3198	
3106 m	3108 sh	3147	3212	
3128 m	3128 m	3219	3285	C_7H stretch

Abbreviations: [a] Calculated with BPW91/6-31 + G*, [b] Calculated with B3LYP/6-31 + G*, Ring 1 = phenyl ring, Ring 2 = isoxazole ring, w = weak, m = medium, s = strong, sh = shoulder, stretch = stretching, bend = bending, twist = twisting, wag = wagging, rock = rocking, s = symmetric, as = asymmetric.

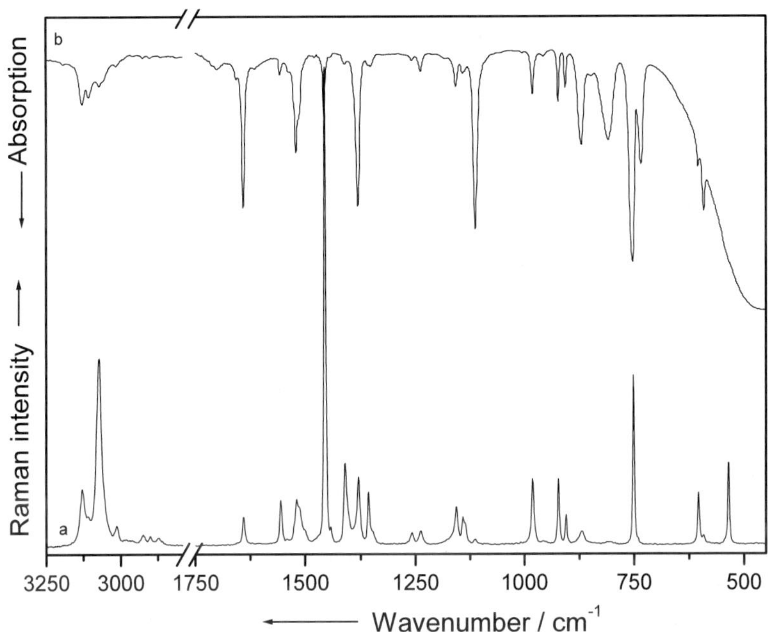

Fig. 3.11 FT-Raman (*a*) and infrared (*b*) spectra of the anthranil molecule. Reprinted with permission from J. Phys. Chem. B 2004, 108, 17491–17496. Copyright 2004 American Chemical Society

3.2.2 Adsorption on the Silver Surface

The SERS spectrum of the anthranil is illustrated in Fig. 3.12 along with the conventional Raman spectrum. Similar to the case of the adsorbed isoxazole molecules (Muniz-Miranda 1999) that exhibit Raman signals only in the presence of chloride anions, SERS enhancements were obtained only for the anthranil species adsorbed on activated silver colloids.

After a close analysis of the spectra (Fig. 3.12) and data summarized in Table 3.5 one can note that the SERS bands are broader and their peak positions and relative intensities are changed relative to their corresponding Raman bands. These spectral features suggest the existence of a strong interaction between the anthranil molecules and the silver surface (Baia et al. 2004), unlike the isoxazole species that shown weak interactions with the metal surface not involving a charge-transfer mechanism (Muniz-Miranda 1999).

Inspection of the structure of the anthranil species suggests that it may bind to the silver surface either through the lone pair electrons of the nitrogen or oxygen atoms or through the π electron system of the aromatic ring (Baia et al. 2004).

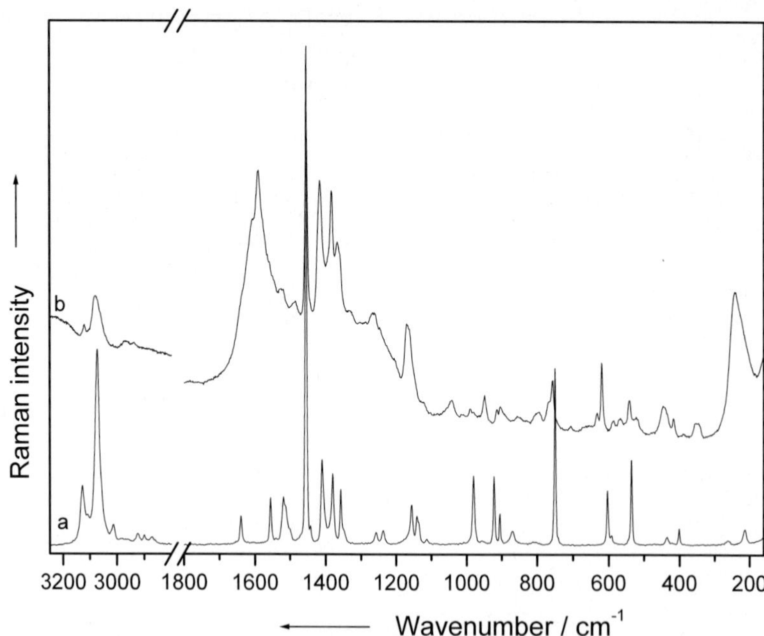

Fig. 3.12 FT-Raman (*a*) and SERS (*b*) spectra of the anthranil molecule. Reprinted with permission from J. Phys. Chem. B 2004, 108, 17491–17496. Copyright 2004 American Chemical Society

In order to determine from a theoretical point of view the most probable binding possibility, DFT calculations have been performed on model compounds having a silver atom bound respectively to the nitrogen and oxygen atoms, and to the phenyl ring. It was found (Baia et al. 2004) that the geometry of the compound with an AgO bond is not stable, and thus the existence of a molecule-metal interaction through the oxygen atom was theoretically excluded. Moreover, the total atomic charges of the nitrogen and oxygen atoms of the free anthranil molecule have been compared, and it was found that at the BPW91 theoretical level their values are –0.307177 for the nitrogen atom and 0.246744 for the oxygen atom, respectively. Considering the fact that an increased negative charge on an atom increases the atom's probability to act as an adsorptive site (Kim et al. 1986) it becomes obvious that the anthranil molecules would bind the colloidal silver particles *via* the nitrogen atoms.

One should also mention that the Ag-anthranil model complex with the AgN bond provides two possible configurations, one with the silver atom below the molecular plane and another one with the silver atom above this plane. According to the results obtained from DFT calculations concerning the total energy of both conformers (Baia et al. 2004), the latter isomer is the most stable one and will be further denoted as the Ag-anthranil model complex with an AgN bond. The optimized geometry of this compound, labelled as complex 1, is illustrated in Fig. 3.13.

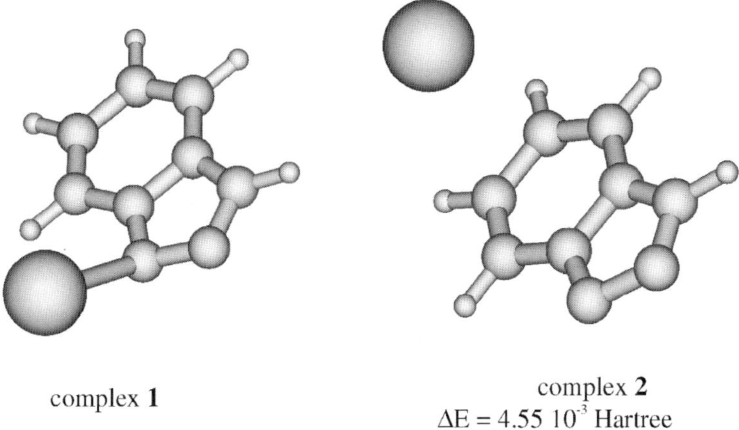

Fig. 3.13 Optimized geometries of the most probable Ag-anthranil SERS complexes. The energy difference between both complexes is also indicated. Reprinted with permission from J. Phys. Chem. B 2004, 108, 17491–17496. Copyright 2004 American Chemical Society

The other adsorption possibility of the anthranil species on the metal surface through the π electron system of the phenyl ring has been also theoretically tested. Thus, calculations on two model complexes, the first one having a silver atom bound to the benzene ring and the second one consisting of two silver atoms, one bound to the phenyl ring and another one bound to the nitrogen atom of the anthranil molecule, have been carried out. The calculations revealed that the geometry of the latter model compound is not stable, and thus this adsorption possibility was excluded. The optimized geometry of the model compound having a silver atom bound to the benzene ring, denoted as complex 2, is shown in Fig. 3.13.

The total energies, including zero point energy corrections, of all model complexes, which were proved to be possible from a theoretical point of view, have been further compared, and it was found that the Ag-anthranil compound having an AgN bond (complex 1) is the most stable one (Baia et al. 2004). The calculated wavenumber values of this compound are given in Table 3.5 and were used for the assignment of the experimentally observed SERS bands.

As was already mentioned, the optimized geometries of the theoretically most probable SERS complexes are shown in Fig. 3.13, while their structural parameters are listed in Table 3.6, together with those of the free anthranil molecule. As can be seen from Table 3.6, the CC bond lengths of the most stable Ag-anthranil complex are shorter than those of the free anthranil molecule, while the NO and NC bond lengths of the isoxazole group are longer. On the contrary, a lengthening of the CC bonds can be observed for the other model SERS complex, while the NO and NC bond lengths of the isoxazole ring are shorter. According to the geometry optimisation calculations, the anthranil unit is planar, while

Table 3.5 Assignment of the normal vibrational modes of the anthranil molecule to the SERS bands

Experimental Raman	SERS	Theoretical Calc.[a]	Vibrational assignment
214 mw	216 sh	208	Ring 1 + 2 out-of-plane def
–	237 s	217	AgN stretch +
264 w	270 sh	257	Ring 1 + 2 out-of-plane def
401 mw	417 m	398	$N_9C_{1,2} + C_{5,6,7}$ bend
435 w	446 m	426	Ring 1 out-of-plane def
536 m	522 m	526	$C_{1,2,3} + C_{4,5,6}$ bend
592 sh	620 m	581	Ring 2 out-of-plane def
603 m	632 w	599	$N_9C_{1,6} + O_8C_{7,6}$ bend
741 sh	758 m	741	CH wag
751 ms	768 sh	752	$C_{1,2,3} + C_{4,5,6} + C_7O_8N_9$ bend
812 vw	794 mw	831	CH twist (ring 1)
869 mw	856 w	864	N_9O_8 stretch
905 m	904 w	888	$C_{1,2,3} + C_{3,4,5} + C_{4,5,6}$ bend
922 m	913 w	912	$C_7O_8N_9$ bend
957 vw	949 m	956	C_7H twist
980 m	990 w	1000	Ring 1 breathing
1114 vw	1120 w	1131	C_7H twist
1140 m	1162 sh	1136	CH bend (ring 1)
1156 m	1170 m	1161	
1237 m	1244 sh	1239	$C_{6,7}O_8$ stretch
1258 m	1265 m	1254	CH rock (ring 1)
1348 sh	1335 m	1345	CH rock (ring 1) + C_1N_9 stretch
1357 m	1367 ms	–	
1380 m	1384 s	1377	$C_{1,6} + C_{3,4}$ stretch
1409 m	1417 s	1403	$C_{1,6} + C_{3,4}$ stretch
1454 vs	1454 vs	1462	$C_{2,3} + C_{4,5}$ stretch
1513 sh	1518 m	1523	$C_{1,2,3}$ s. stretch
1518 m	1524 m	–	
1555 m	1589 s	1563	$C_{4,5,6}$ s. stretch
1640 m	1640 sh	1639	$C_{1,2,3} + C_{4,5,6} + C_{5,6,7}$ as. stretch
3071 s	3078 ms	3130	CH stretch (ring 1)
3128 m	3120 w	3226	C_7H stretch

Abbreviations: [a] Calculated on the complex 1 with: BPW91/6-31 + G* for N, C, O, H atoms and LanL2DZ for Ag atom, Ring 1 = phenyl ring, Ring 2 = isoxazole ring, w = weak, m = medium, s = strong, sh = shoulder, stretch = stretching, bend = bending, twist = twisting, wag = wagging, rock = rocking, s = symmetric, as = asymmetric
Reprinted with permission from J. Phys. Chem. B 2004, 108, 17491-17496. Copyright 2004 American Chemical Society

a slight deviation from the planarity of this unit was observed for both SERS complexes (Baia et al. 2004). The calculated dihedral angles between the benzene and isoxazole rings of both compounds were found to be 0.487 and 0.345 degrees, respectively. These changes of the structural parameters of the SERS complexes,

Table 3.6 Selected theoretical structural parameters of two Ag-anthranil model complexes compared with those of the free anthranil molecule

	Complex 1[a]	Complex 2[a]	Anthranil[b]
Bond lengths ($\mathring{A}$)			
$(CC_{Ph})_{average}$	1.416	1.419	1.418
C_1C_2	1.425	1.430	1.429
C_2C_3	1.381	1.380	1.381
C_3C_4	1.436	1.445	1.439
C_4C_5	1.380	1.385	1.381
C_5C_6	1.426	1.425	1.427
C_1C_6	1.447	1.450	1.449
C_6C_7	1.384	1.384	1.385
C_7O_8	1.347	1.348	1.349
O_8N_9	1.410	1.407	1.408
N_9C_1	1.347	1.340	1.341
Bond angles (degree)			
$C_1N_9O_8$	104.756	104.132	104.205
$N_9O_8C_7$	110.289	110.908	110.842
$O_8C_7C_6$	110.101	109.793	109.743
$C_2C_1N_9$	127.893	127.199	127.287
$C_5C_6C_7$	136.192	136.528	136.557
Dihedral angle	179.513	179.655	180.0

Abbreviations: complex 1 = Ag-anthranil model complex having an Ag-N bond, complex 2 = Ag-anthranil model complex having an Ag atom bound to the phenyl group. [a] Calculated with BPW91/6-31 + G* for the N, C, O, H atoms and LanL2DZ for Ag atom, [b] Calculated with BPW91/6-31 + G*. Reprinted with permission from J. Phys. Chem. B 2004, 108, 17491–17496. Copyright 2004 American Chemical Society

relative to those of the free anthranil molecule, should be mirrored in their vibrational spectra.

The SERS spectrum of anthranil (Fig. 3.12) presents an intense band at 237 cm^{-1}, which is characteristic for the SERS spectra of N-adsorbed species and is ascribed to the AgN stretching vibration (Chowdhury et al. 2000). The existence of this band can be regarded as evidence of the anthranil species bonding to the silver surface through the lone pair electrons of the nitrogen atom, even if out-of-plane deformation vibrations of both benzene and isoxazole rings give rise to bands in approximately the same spectral region.

As one can see from Fig. 3.12 and Table 3.5, the experimental SERS bands at 417, 632, and 768 cm^{-1} attributed to the in-plane deformation vibration of both rings are shifted to higher wavenumbers up to 20 cm^{-1} relative to their corresponding Raman bands. The band at 990 cm^{-1} due to the breathing vibration of the benzene ring is also blue shifted in the SERS spectrum by 10 cm^{-1}. The SERS band at 856 cm^{-1} given by the NO stretching vibration is shifted to lower wavenumbers by

$13\,\mathrm{cm}^{-1}$ in comparison with its analogue Raman band. As revealed by theoretical calculations the NO bond length of the most stable SERS complex is longer than that of the free molecule. The behavior of the NO stretching band in the SERS spectrum supports the theoretical predictions that the Ag-anthranil complex with an AgN bond is the most probable one, otherwise a shift to higher wavenumber values of this band should have been observed in the SERS spectrum. From Fig. 3.12 and Table 3.5 one can also remark that most of the SERS bands present in the spectral range between 1400 and $1600\,\mathrm{cm}^{-1}$, which are attributed to CC stretching vibrations, are shifted towards higher wavenumbers, relative to their analogue Raman bands. This behavior agrees with the geometrical changes evidenced from the comparison of the CC bond lengths of the most stable SERS complex with those of the free anthranil molecule, and confirms again that the model complex having an AgN bond is the most probable one. Thus, by taking into account the predictions of DFT calculations carried out on different SERS model complexes and from the comparison of the corresponding SERS and Raman bands, one can conclude (Baia et al. 2004) that the anthranil molecules are adsorbed on the colloidal silver particles through the nonbonding electrons of the nitrogen atoms.

For a better understanding of the adsorption behavior of the anthranil species on the metal surface the electronic absorption spectra of pure silver sol and mixture of activated silver colloid and anthranil solution have been recorded and are presented in Fig. 3.14.

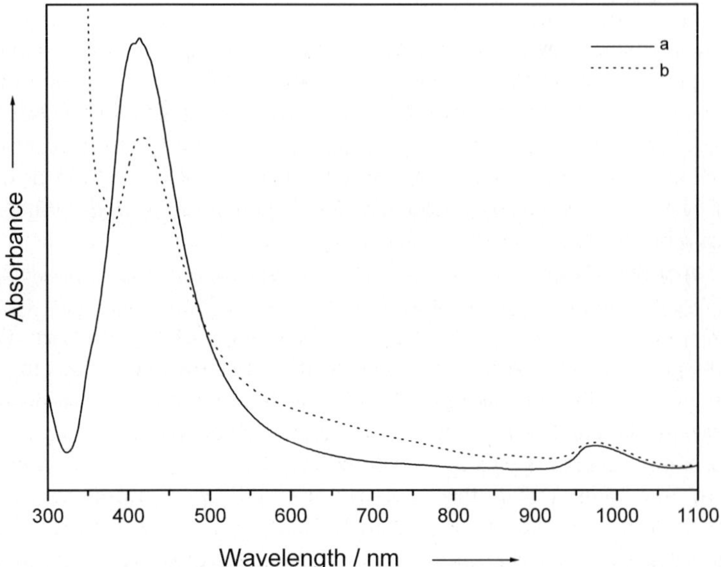

Fig. 3.14 Absorption spectra of pure silver colloid (*a*), and with 10^{-1} M anthranil and 10^{-2} M NaCl (*b*). Reprinted with permission from J. Phys. Chem. B 2004, 108, 17491–17496. Copyright 2004 American Chemical Society

The spectrum of pure silver colloid (Fig. 3.14*a*) shows a single absorption maximum at 412 nm, due to small particle plasma resonance. The addition of anthranil to the activated sol (Fig. 3.14*b*) causes a significant decrease and a shift of this absorption band towards longer wavelengths by 6 nm, while a new broad absorption signal appears at longer wavelength values (around 700 nm). When two metallic spheres approach each other (Muniz-Miranda 1999) this band approximately remains at the original single spheres wavelength, while a new band occurs at longer wavelengths. The latter absorption peak is known to arise from the aggregates of colloid particles formed upon addition of the adsorbed molecules (Blatchford et al. 1982). However, this band has been reported to be due to a charge transfer between the metal and the adsorbate (Sanchez-Cortes et al. 1995). Unlike the isoxazole molecules, whose heteroatoms strongly impair the π electron delocalisation and are considered to be physisorbed on the silver surface (Muniz-Miranda 1999), the major changes evidenced between the Raman and SERS spectra of anthranil corroborated with DFT calculations data obtained from different Ag-anthranil model complexes, and the features of the absorption spectrum of activated silver sol with added adsorbens allow us to conclude the chemisorption of the anthranil species on the colloidal silver particles, and the contribution of the charge-transfer effect to the overall SERS enhancement.

By inspection of specific enhanced bands in the SERS spectrum in agreement with the surface selection rules (Hallmark and Campion 1986, Moskovits and Suh 1984, Moskovits and DiLella 1980) the orientation of the adsorbed species relative to the metal surface can be determined. According to these rules, the vibrational modes that involve a large change of the polarizability perpendicular to the metal surface are the most enhanced. By comparing the SERS and Raman spectra of the anthranil molecule, obviously the bands attributed to out-of-plane vibrations are mainly enhanced. Thus, the SERS bands at 446 and 620 cm^{-1} assigned to the out-of-plane deformation vibrations of the phenyl and isoxazole rings are enhanced in comparison with their corresponding Raman bands, while the bands given by the in-plane ring deformation vibrations present in the SERS spectrum at 632 and 768 cm^{-1} are only weakly enhanced. The bands evidenced at 758, 794, 949, and 1120 cm^{-1} in the SERS spectrum and assigned to the out-of-plane deformation vibrations of the CH groups of both rings (see Table 3.5) are also enhanced, compared to their analogue Raman bands. However, in the high wavenumber region of the SERS spectrum, one can observe the enhancement of the bands at 3078 and 3120 cm^{-1}, which were clearly attributed to the CH stretching vibrations. Moreover, the band at 990 cm^{-1} assigned to the breathing vibration of the phenyl ring together with the bands due to the CC stretching vibrations of both rings situated in the spectral range between 1400 and 1600 cm^{-1} are also enhanced in the SERS spectrum. Having in view all these features of the SERS spectrum one can predict (Baia et al. 2004) that the adsorbed anthranil molecules adopt a tilted orientation relative to the silver surface. By assuming that the silver atom of the most stable Ag-anthranil model complex approximates the metallic surface and applying the surface selection rules to this compound the bands enhancement evidenced in the experimental SERS spectrum should appear. Therefore, it can be affirmed (Baia et al. 2004) that the orientation of the adsorbed species determined by following the enhancement of the experimental SERS bands according to the surface selection rules consistently agrees with that of the Ag-anthranil model complex. The theoretical value of the dihedral angle formed by the silver atom and the isoxazole plane of the molecule was found to be 137.882 degrees.

3.2.3 Conclusions

The Raman and infrared spectra of the anthranil molecule have been recorded and the assignment of the vibrational modes has been performed on the basis of the results of DFT calculations. The SERS spectrum of the molecule in colloidal silver suspension has been also recorded and analyzed.

By correlating the spectroscopic changes evidenced between the Raman and SERS spectra, and the results of DFT calculations performed on different Ag-anthranil model complexes, it was concluded that the anthranil molecule is adsorbed on the colloidal silver surface through the lone pair electrons of the nitrogen atom. The contribution of the charge-transfer effect to the overall SERS enhancement has been confirmed by the spectral modifications of the electronic absorption spectrum of activated silver sol with added adsorbate. By following the enhancement of the SERS bands according to the surface selection rules, a titled orientation of the adsorbed anthranil species relative to the metal surface has been predicted.

References

Alkalis SA, Beck G, Grätzel M (1975) Laser photoionization of phenothiazine in alcoholic and aqueous micellar solution. Electron transfer from triplet states to metal ion acceptors. J Am Chem Soc 97:5723–5729

Anderson MR, Evans DH (1988) Surface-enhanced Raman study of the effect of electrode potential and solution pH upon the interfacial behavior of 4-pyridinecarboxaldehyde. J Am Chem Soc 110:6612–6617

Armuth V, Berenblum I (1982) A possible in vivo skin model for tumour promoter assays. Cancer Lett 15:343–346

Baia M, Baia L, Kiefer W, Popp J (2004) Surface-enhanced Raman scattering and density functional theoretical study of anthranil adsorbed on colloidal silver particles. J Phys Chem B 108:17491–17496

Bolboaca M, Iliescu T, Kiefer W (2004) Infrared absorption, Raman, and SERS investigations in conjunction with theoretical simulations on a phenothiazine derivative. Chem Phys 298:87–95

Bohn W, Rutter G, Hohenberg H, Mannweiler K (1983) Inhibition of measles virus budding by phenothiazines. Virology 130:44–55

Bolboaca M, Iliescu T, Paizs C, Irimie FD, Kiefer W (2003) Raman, infrared, and surface-enhanced Raman spectroscopy in combination with ab initio and density functional theory calculations on 10-isopropyl-10H-phenothiazine-5-oxide. J Phys Chem A 107:1811–1818

Blatchford CG, Campbell JR, Creighton JA (1982) Plasma resonance-enhanced Raman scattering by adsorbates on gold colloids: the effects of aggregation. Surf Sci 120:435–455

Campion A, Kambhampati P (1998) Surface enhanced Raman scattering. Chem Soc Rev 27:241–250

Candurra NA, Maskin L; Damonte EB (1996) Inhibition of arenavirus multiplication in vitro by phenothiazines. Antiviral Res 31:149–158

Chetty M, Pillay VL, Moodley SV, Miller R (1996) Response in chronic schizophrenia correlated with chlorpromazine, 7-OH-chlorpromazine and chlorpromazine sulfoxide levels. Eur Neuropsychopharmacol 6:85–91

Chowdhury J, Ghosh M, Misra TN (2000) pH-dependent surface enhanced Raman scattering of 8-hydroxy quinoline adsorbed on silver hydrosol. J Colloid Interface Sci 228:372–378

Chowdhury J, Ghosh M, Misra TN (2000) Surface enhanced Raman scattering of 2,2′ biquinoline adsorbed on colloidal silver particles. Spectrochim Acta A 56:2107–2115

Creighton JA (1983) Surface Raman electromagnetic enhancement factors for molecules at the surface of small isolated metal spheres: The determination of adsorbate orientation from SERS relative intensities. Surf Sci 124:209–219

Dastidar SG, Ganguly K, Chaudhuri K, Chakrabarty AN (2000) The anti-bacterial action of diclofenac shown by inhibition of DNA synthesis. Int J Antimicrobial Agents 14:249–251

Delay J, Deniker P, Harl JM (1952) Utilisation en thérapeutique psychiatrique d'une phénothiazine d'action centrale elective. Ann Med Psychol 2: 111–117

Doppler T, Schmid H, Hansen HJ (1979) Zur Photochemie von 2, 1-Benzisoxazolen (Anthranilen) und thermischen und photochemischen Umsetzungen von 2-Azido-acylbenzolen in stark saurer Lösung. Helv Chimica Acta 62:271–303

Dryhurst CG (1977) Electrochemistry of biological molecules. Academic Press, New York

Ford JM, Prozialeck WC, Hait WN (1989) Structural features determining activity of phenothiazines and related drugs for inhibition of cell growth and reversal of multidrug resistance. Mol Pharmacol 35:105–115

Fu S, Zhang P (1992) Chemical effect of chloride ions on SERS in silver sol. J Raman Spectrosc, 23:93–97

Gao P, Weaver MJ (1985) Surface-enhanced Raman spectroscopy as a probe of adsorbate-surface bonding: Benzene and monosubstituted benzenes adsorbed at gold electrodes. J Phys Chem 89:5040–5046

Gao X, Davies JP, Weaver MJ (1990) A test of surface selection rules for surface-enhanced Raman scattering: the orientation of adsorbed benzene and monosubstituted benzenes on gold. J Phys Chem 94:6858–6865

Hallmark VM, Campion A (1986) Selection rules for surface Raman spectroscopy: experimental results. J Chem Phys 84:2933–2941

Hehre WJ, Radom L, Schleyer PVR, Pople JA (1986) Ab initio molecular orbital theory. Wiley, New York

Henry BR, Kasha M (1965) Triplet–triplet absorption studies on aromatic and heterocyclic molecules, at 77 K. J Chem Phys 47:3319–3327

Hester RE, Williams KPJ (1981) Free radical studies by resonance Raman spectroscopy: phenothiazine, 10-methylphenothiazine, and phenoxazine radical cations. J Chem Soc Perkin Trans 2: 852–858

He YZ, Cui JP, Mallard WG, Tsang W (1988) Homogeneous gas-phase formation and destruction of anthranil from o-nitrotoluene decomposition. J Am Chem Soc 110:3754–3759

Hildebrandt P, Keller S, Hoffmann A, Vanhecke F, Schrader B (1993) Enhancement factor of surface-enhanced Raman scattering on silver and gold surfaces upon near-infrared excitation. Indication of an unusual strong contribution of the chemical effect. J Raman Spectrosc 24:791–796

Kim SK, Joo TH, Suh SW, Kim MS (1986) Surface-enhanced Raman scattering (SERS) of nucleic acid components in silver sol: Adenine series. J Raman Spectrosc 17:381–386

Lombardi JR, Birke RL, Lu T, Xu J (1986) Charge transfer theory of surface enhanced Raman spectroscopy; Herzberg Teller contributions. J Chem Phys 84:4174–4180

Liang EJ, Engert C, Kiefer W (1993) Surface-enhanced Raman scattering of pyridine in silver colloids excited in the near-infrared region. J Raman Spectrosc 24:775–779

Manolov P, Todorov S (1974) Experimental studies on Dolyspan regarding its antiulcerous action. Eksp Med Morfol 13:36–39

Mille G, Guiliano M, Angelelli JM, Chouteau J (1980) Benzisoxazoles -1,2 et -2,1: Analyse vibrationnelle infrarouge et Raman. J Raman Spectrosc 9:339–343

McDowell JJH (1976) The crystal and molecular structure of phenothiazine. Acta Crystallogr B5:10–20

Moskovits M (1985) Surface-enhanced spectroscopy. Rev Mod Phys 57:783–826

Moskovits M, DiLella DP (1980) Surface-enhanced Raman spectroscopy of benzene and benzene-d6 adsorbed on silver. J Chem Phys 73:6068–6075

Moskovits M, Suh JS (1984) Surface selection rules for surface enhanced Raman spectroscopy: calculations and applications to surface-enhanced Raman spectrum of phthalazine on silver. J Phys Chem 88:5526–5530

Motohashi N (1991) Phenothiazines and calmodulin. Anticancer Res 11:1125–1164

Motohashi N, Sakagami H, Kurihara T, Csuri K, Molnar J (1992) Antiplasmid activity of phenothiazines, benzo[a]phenothiazines and benz[c]acridines. Anticancer Res 12:135–139

Muniz-Miranda M (1999) SERS investigation on five-membered heterocyclic compounds: isoxazole, oxazole and thiazole. Vib Spectrosc 19:227–232

Pan D, Phillips DL (1999) Raman and density functional study of the S0 state of phenothiazine and the radical cation of phenothiazine. J Phys Chem A 103:4737–4743

Rauhut G, Pulay P (1995) Transferable scaling factors for density functional derived vibrational force fields. J Phys Chem 99:3093–310026

Sanchez-Cortez S, Garcia-Ramos JV (1992) SERS of cytosine and its methylated derivatives on metal colloids. J Raman Spectrosc 23:61–66

Sanchez-Cortes S, Garcia-Ramos JV, Morcillo G, Tinti A (1995) Morphological study of silver colloids employed in surface-enhanced Raman spectroscopy: activation when exciting in visible and near-infrared regions. J Colloid Interface Sci 175:358–368

Scott AP, Radom L (1996) Harmonic vibrational frequencies: an evaluation of Hartree-Fock, Møller-Plesset, quadratic configuration interaction, density functional theory, and semiempirical scale factors. J Phys Chem 100:16502–16513

Shine HJ, Mach EE (1965) Ion radicals. V. Phenothiazine, phenothiazine 5-oxide, and phenothiazone-3 in acid solutions. J Org Chem 30:2130–2139

Sun SC, Bernard I, Birke RL, Lombardi JR (1985) The effect of pH, chloride ion and background electrolyte concentration on the SERS of acidified pyridine solutions. J Electroanal Chem 196:359–374

Tosa M, Paizs C, Majdik C, Poppe L, Kolonits P, Silberg IA, Novák L, Irimie FD (2001) Selective oxidaton methods for preparation of N-alkylphenothiazine sulfoxides and sulfones. Heterocyclic Commun 7:277–282

Takahashi M, Furukawa H, Fujita M, Ito M (1987) Surface-enhanced Raman spectra of phthalazine: anion-induced reorientation on a silver electrode. J Phys Chem 91:5940–5943

Wong MW (1996) Vibrational frequency using density functional theory. Chem Phys Lett 256:391–399

4 Anti-Inflammatory Drugs

4.1 Diclofenac Sodium

Diclofenac sodium (DCFNa) is a sodium salt of an aminophenyl acetic acid (see Fig. 4.1) and is a well-known representative of nonsteroidal anti-inflammatory drugs (NSAIDs) (Tunçay et al. 2000, Kovala-Demertzi et al. 1993). Like other NSAIDs, diclofenac sodium is clinically prescribed as an antipyretic, analgesic, and anti-inflammatory agent (Abdel-Hamid et al. 2001, Todd and Sorkin 1988, Moser et al. 1990). The antipyretic effect is due to a resetting of the hypothalamic temperature-regulating center, whereas the anti-inflammatory and analgesic effects are due to the inhibition of prostaglandin synthesis (Abdel-Hamid et al. 2001, Todd and Sorkin 1988, Moser et al. 1990). Therapeutically, NSAIDs are indicated

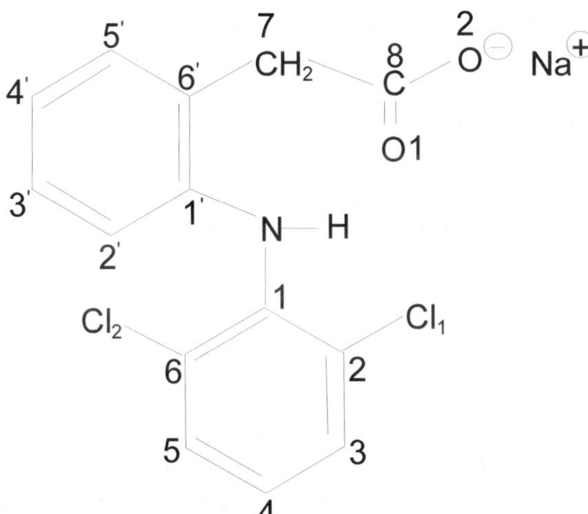

Fig. 4.1 Schematic structure of DCFNa. Reprinted from Chem. Phys. 298, Iliescu T, Baia M, Kiefer W, FT-Raman, surface-enhanced Raman spectroscopy and theoretical investigations of diclofenac sodium, 167–174, copyright 2004 with permission from Elsevier

to control pain and inflammation of rheumatic and non-rheumatic origin. In an intensive search for antimicrobial action among the NSAIDs, DCFNa exhibited significant potential against both Gram-positive and Gram-negative bacteria. Furthermore, it was shown that DCFNa demonstrated clearance of the challenged pathogenic bacteria from the liver and the spleen. More recently, it was illustrated that the mechanism of antibacterial action of diclofenac is by inhibition of bacterial DNA synthesis (Dastidar et al. 2000).

Knowledge of the structure of the DCFNa molecule is essential to understand its pharmaceutical action. Many spectroscopic and non-spectroscopic techniques were used to study this molecular species. The FT-infrared spectrum of DCFNa was obtained and analyzed by Szejtli (Szejtli 1982) and Kovala-Demertzi et al. (Kovala-Demertzi et al. 1993). Other methods like calorimetry (Sastry et al. 1987, Sastry et al. 1989), UV spectrophotometry (Arrawal et al. 1988), gas (Hennig et al. 1987, Schneider and Degen 1981), liquid chromatography (Grandjean et al. 1989, Godbilon et al. 1985), and nuclear magnetic resonance spectroscopy (Abdel Fattah et al. 1988) were used to study DCFNa molecular structure.

DCFNa has limited water solubility, especially in gastric juice, and is unstable in an aqueous solution (Pose-Vilarnovo et al. 1999). This limited solubility in an acidic medium engenders problems in its oral bioavailability and it is a drawback in terms of its formulation in controlled release devices. A possibility to overcome these limitations is the complexation of DCFNa with β-cyclodextrin (βCD) (Pose-Vilarnovo et al. 1999). βCD is a cyclic oligosaccharide consisting of seven glucopyranose units that can be represented as a truncated cone structure with the wide and narrow rims occupied by the secondary and primary hydroxyl group, respectively (see Fig. 4.2). The central cavities of these molecules (host molecules) are hydrophobic, and thus are able to encapsulate a wide variety of molecules (guest molecules) such as: acids, ions, halides, aliphatic molecules, alicyclic molecules, and aromatic hydrocarbons (Amado et al. 1994, Amado et al. 2000), based on the physical fit and the chemical affinity.

The interaction of DCFNa with βCD has been studied both in a solution (Pose-Vilarnovo et al. 1999, Astilean et al. 1997, Whittaker et al. 1996, Mucci et al. 1999) and in the solid state (Caira et al. 1994, Bratu et al. 1998, Cwiertnia et al. 1999) by using different experimental techniques like nuclear magnetic resonance spectroscopy, infrared absorption spectroscopy, and X-ray diffraction and, depending on the aggregation state and the preparation method of the DCFNa-βCD complex, different inclusion ways of the guest molecule into the βCD cavities have been reported.

Some authors have reported the existence in solution of two isomeric 1 : 1 DCFNa-βCD complexes having either the dichlorophenyl group or the phenylacetate ring included in the βCD cavity (Astilean et al. 1997, Whittaker et al. 1996). The possibility to obtain the 1 : 2 guest-host complex, which imply the inclusion of both rings of the DCFNa molecule into two βCD entities was also indicated (Pose-Vilarnovo et al. 1999). Moreover, A. Mucci and her coworkers (Mucci et al. 1999) have evidenced by means of nuclear magnetic resonance techniques the existence of multiple equilibriums involving both 1 : 1 and 2 : 1 guest-host species.

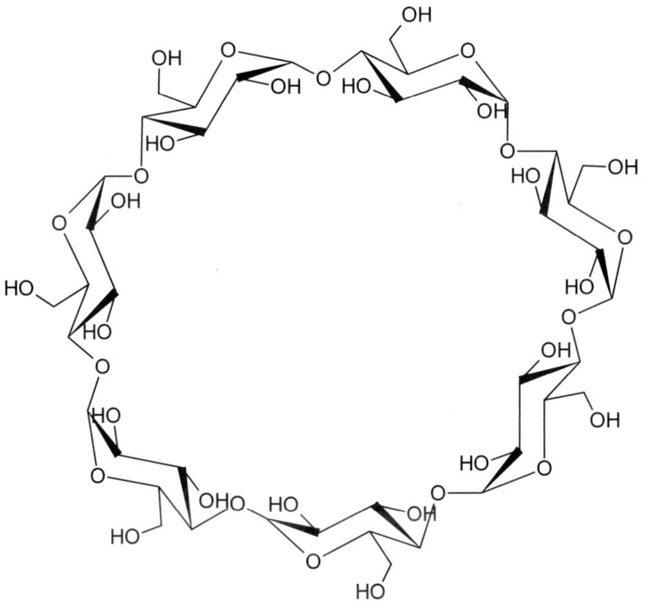

Fig. 4.2 Schematic drawing of the βCD molecule. Reprinted from Eur. J. Pharma. Sci. 22, Iliescu T, Baia M, Miclaus V, A Raman spectroscopic study of the diclofenac sodium-β-cyclodextrin interaction, 487–495, copyright 2004 with permission from Elsevier

On the other hand, the studies carried out on the DCFNa-βCD inclusion complex in the solid state have revealed the existence of 1 : 1 species having the βCD cavities preferentially filled by the phenylacetate rings (Caira et al. 1994, Bratu et al. 1998) or by acetate groups interacting with the host molecules (Cwiertnia et al. 1999).

In order to elucidate the interaction between the DCFNa and βCD in the solid state complex, the analysis of the free DCFNa molecule becomes obvious. Thus, a relatively detailed experimental and theoretical study of the DCFNa molecule has been performed and is presented in the next sections. The first part of the study is focused on the investigation of the DCFNa molecule from an analytical (FT-Raman spectroscopy) and theoretical (DFT and ab initio calculations) point of view. SERS spectra of DCFNa in silver sol at different pH values were also recorded and analyzed in order to find out the adsorption behavior of the molecules on colloidal silver particles (reprinted from Chem. Phys. 298, Iliescu T, Baia M, Kiefer W, FT-Raman, surface-enhanced Raman spectroscopy and theoretical investigations of diclofenac sodium, 167–174, copyright 2004 with permission from Elsevier).

4.1.1 Vibrational Analysis

Due to the flexibility of the acetate group, the DCFNa molecule allows for several conformers. Ab initio and DFT calculations have been performed at the

RHF/6-31G*, BPW91/6-31G*, and B3LYP/6-31G* levels of theory on two of the most probable conformers in order to find the most stable one (Iliescu et al. 2004a, Iliescu et al. 2003). The optimized geometries of these two conformers calculated at the BPW91/6-31G* theoretical level are illustrated in Fig. 4.3. Analytical harmonic vibrational modes have also been calculated in order to ensure that the optimized structures correspond to minima on the potential energy surface. The calculations performed on both isomers at all theoretical levels demonstrate (Iliescu et al. 2004a), in agreement with the experimental data obtained from X-ray diffraction experiments on tetrahydrate DCFNa crystals (Reck et al. 1988) that the conformer 2 is energetically more stable by an energy difference of approximately 18 kJ/mol (RHF), 30 kJ/mol (BPW91), and 27 kJ/mol (B3LYP), respectively.

Single crystal X-ray diffraction studies of tetrahydrate DCFNa (Reck et al. 1988) indicate that the dihedral angle between the two phenyl rings has a value of

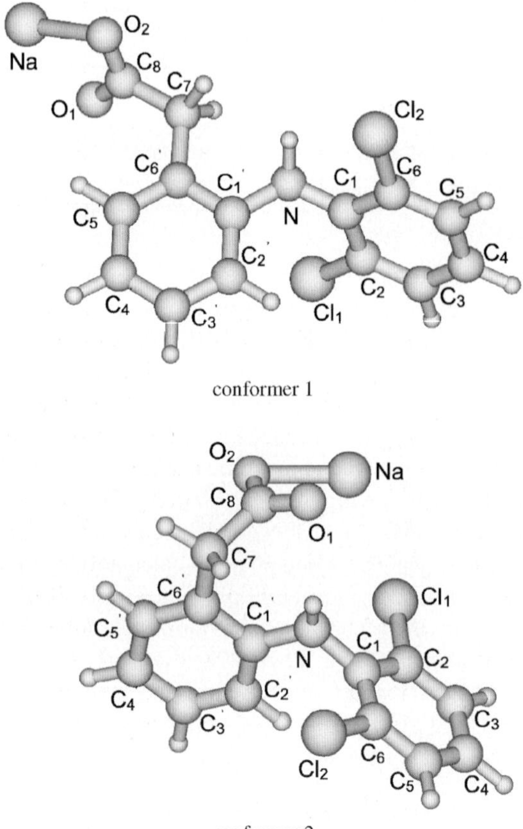

Fig. 4.3 Optimized geometries of two conformational isomers of the DCFNa molecule obtained at the BPW91/6-31G* level of theory. Reprinted from Chem. Phys. 298, Iliescu T, Baia M, Kiefer W, FT-Raman, surface-enhanced Raman spectroscopy and theoretical investigations of diclofenac sodium, 167–174, copyright 2004 with permission from Elsevier

58.3°, smaller than that of the free acid, found to be 69°. The crystallographic investigations revealed also the existence of hydrogen bonds between the potential NH donor center and the acceptor oxygen atom of the carbonyl group in the crystal structure of free diclofenac acid molecules, while in the tetrahydrate DCFNa crystals such interactions were not evidenced (Iliescu et al. 2004a).

Selected optimized structural parameters of DCFNa calculated by various methods (Iliescu et al. 2004a, Iliescu et al. 2003) are given in Table 4.1, along with the available X-ray values (Reck et al. 1988).

Table 4.1 Selected calculated bond lengths (Å) and angles (degree) of DCFNa compared to the experimental data

	Calc.[a]	Calc.[b]	Calc.[c]	Exp.[d]
Bond lengths (Å)				
C_1-$C_{2average}$	1.386	1.397	1.399	1.394
C_2-Cl_1	1.751	1.763	1.765	1.753
C_6-Cl_2	1.736	1.745	1.749	1.742
C_1-N	1.406	1.391	1.396	1.396
$C_{1'}$-N	1.415	1.413	1.416	1.422
$C_{1'}$-$C_{2'average}$	1.388	1.386	1.390	1.377
$C_{6'}$-C_7	1.570	1.599	1.611	1.608
C_7-C_8	1.529	1.530	1.532	1.521
C_8-O_1	1.239	1.240	1.244	1.236
C_8-O_2	1.248	1.282	1.284	1.280
O_2-Na	2.208	2.214	2.183	2.3 ÷ 3
O_1-Na	2.234	2.224	2.226	2.3 ÷ 3
Angles (degree)				
dihedral	57.954	58.257	56.717	58.3
C_1-C_2-$C_{3average}$	119.998	120.990	121.995	122.6
C_1-C_2-Cl_1	119.553	119.122	120.004	120.1
C_1-C_6-Cl_2	119.937	119.760	119.358	120.4
C_1-N-$C_{1'}$	121.827	125.013	124.525	124.2
$C_{1'}$-$C_{2'}$-$C_{3'average}$	119.999	119.996	119.998	119.98
$C_{1'}$-$C_{6'}$-C_7	122.234	121.386	121.949	120.5
$C_{6'}$-C_7-C_8	112.859	113.245	112.920	113.1
C_7-C_8-O_1	119.746	121.372	121.712	122.4
C_7-C_8-O_2	115.622	115.074	114.615	113.8
O_1-C_8-O_2	124.315	123.552	123.486	123.7

Abbreviations: [a]Calculated with RHF/6-31G*, [b]Calculated with BPW91/6-31G*, [c]Calculated with B3LYP/6-31G*, [d]Ref. (Reck et al. 1988). Reprinted from Chem. Phys. 298, Iliescu T, Baia M, Kiefer W, FT-Raman, surface-enhanced Raman spectroscopy and theoretical investigations of diclofenac sodium, 167–174, copyright 2004 with permission from Elsevier

As it can be observed, the theoretical values of the dihedral angle between the two phenyl rings of the DCFNa reproduce well the experimental one and are smaller compared to the dihedral angle of the free acid. According to the crystallographic analysis the sodium atoms are surrounded by oxygen atoms belonging either to water molecules or to carboxylate groups, the NaO bond lengths having values in the range between 2.3 and 3 Å. As one can see from Table 4.1, the unscaled calculated bond lengths and bond angles agree with the reported parameters (Reck et al. 1988), the observed differences being most probably due to the intermolecular interactions, which occur in the crystal between the diclofenac anion, the sodium cation, and water molecules. Moreover, the theoretical calculations were performed for the gas phase, while the experimental data are for the solid phase.

The FT-Raman spectrum of DCFNa in the range from 3200 to $100\,cm^{-1}$ with the calculated unscaled Raman intensities are illustrated in Fig. 4.4. The observed Raman bands with their vibrational assignment accomplished with the help of theoretical calculations are presented in Table 4.2.

It is known (Hehre et al. 1986) that ab initio harmonic vibrational wavenumbers are larger than the fundamentals observed experimentally, most probably due to the neglect of the anharmonicity effects in the theoretical treatment. The incomplete incorporation of electron correlation and the use of finite basis sets contribute also to the disagreement. However, the overestimation of ab initio

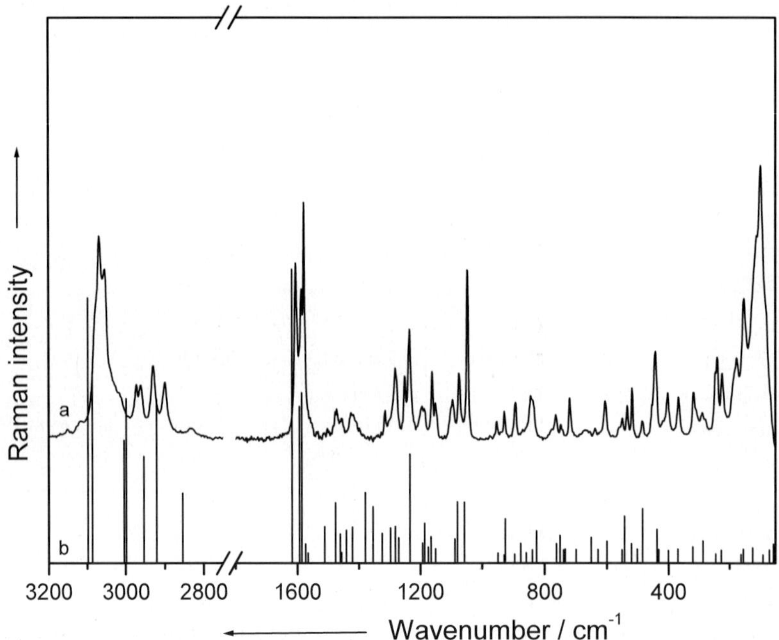

Fig. 4.4 FT-Raman spectrum of polycrystalline DCFNa (*a*) and the calculated Raman wavenumbers (BPW91/6-31G*) (*b*). Reprinted from Chem. Phys. 298, Iliescu T, Baia M, Kiefer W, FT-Raman, surface-enhanced Raman spectroscopy and theoretical investigations of diclofenac sodium, 167–174, copyright 2004 with permission from Elsevier

harmonic vibrational wavenumbers is found to be relatively uniform, therefore scaling factors are often applied to adjust the experimental values (Scott and Radom 1996). The RHF calculated vibrational wavenumbers presented in Table 4.2 have been uniformly scaled by a factor of 0.8953, according to the work of Scott and Radom (Scott and Radom 1996). Even after scaling, in comparison to the experiment, the RHF wavenumbers are overestimated in the high wavenumber region, but are comparable to the experimental values in the low wavenumber region.

Table 4.2 Assignment of the theoretical wavenumber values (cm^{-1}) to the experimental FT-Raman bands of the DCFNa molecule. Selected infrared bands are also presented

Infrared	Raman	Calc.[a]	Calc.[b]	Calc.[c]	Vibrational assignment
–	152 ms	143	149	147	Ring 2 out-of-plane def
–	175 ms	183	161	178	CONa def + $C_{6',7,8}$ def + C_7H_2 def
–	223 m	210	217	218	Ring 2 out-of-plane def
–	239 m	231	240	242	CONa def + C_7H_2 rock
–	288 m	291	277	299	Ring 1 out-of-plane def + $C_1NC_{1'}$ def
–	318 m	321	334	323	$C_1NC_{1'}$ def + $C_{6',7,8}$ def + CCl def
–	367 m	354	356	357	
–	402 m	394	397	391	$O_1C_8O_2$ def + $Cl_1C_{2,3}$, $Cl_2C_{6,5}$ def + C_7H_2 def
–	417 sh	429	425	423	Ring 2 out-of-plane def + NaO stretch
–	442 ms	442	446	443	
–	484 m	483	476	470	NH def
–	517 m	518	524	522	Ring 2 + 1 out-of-plane def
–	533 m	534	537	524	
–	549 m	540	539	536	$C_{6',7,8}$ def + C_7H_2 def
–	604 m	587	599	589	Ring 1 + 2 in-plane def
–	637 w	645	630	621	$O_1C_8O_2$ def + C_7H_2 def
–	653 w	655	644	647	$C_1NC_{1'}$ def + ring 1 out-of-plane def
716 w	718 w	722	724	731	CH wag (ring 2 + 1)
748 s	747 m	753	748	741	
769 mw	763 m	760	762	759	C-Cl stretch + ring 1 + 2 in-plane def
–	774 sh	763	766	768	
838 w	844 m	838	848	837	CH twist (ring 2 + 1)
868 w	868 w	861	866	867	
891 vw	892 m	877	894	893	
925 vw	928 m	927	927	929	$C_{7,8}$ stretch
935 vw	940 sh	940	934	936	CH twist (ring 2 + C_7H_2)
950 w	952 m	962	954	956	
1045 w	1046 s	1059	1051	1050	Ring 2 breathing
1073 vw	1073 ms	1074	1081	1066	Ring 1 breathing

Table 4.2 (Continued)

Infrared	Raman	Calc.[a]	Calc.[b]	Calc.[c]	Vibrational assignment
1090 w	1094 m	1094	1094	1085	C_7H_2 wag + CH bend (ring 1 + 2)
1153 w	1150 m	1151	1149	1150	CH bend (ring 1 + 2)
1161 w	1160 ms	1158	1167	1162	
1181 vw	1186 m	1191	1188	1187	CH rock (ring 1)
1200 m	1193 m	1196	1197	1197	C_7H_2 twist
1236 vw	1235 s	1213	1243	1221	$C_1NC_{1'}$ s. stretch + CH rock (ring 1 + 2) + C_7H_2 wag
1248 w	1250 m	1260	1269	1250	$C_{6',7}$ stretch + CH rock (ring 1 + 2)
1284 m	1281 ms	1272	1284	1277	CH rock (ring 1 + 2) + C_7H_2 wag
1306 s	1303 sh	1298	1307	1296	$C_1NC_{1'}$ as. stretch + CH rock (ring 1 + 2) + C_7H_2 wag
–	1327 vw	1316	1320	1311	Ring 1 stretch + C_7H_2 wag
–	1360 vw	1358	1357	1363	Ring 2 stretch + C_7H_2 wag
1402 s	1398 vw	1404	1385	1406	$C_{7,8}$ stretch + $O_1C_8O_2$ s. stretch
–	1417 m	1433	1426	1437	CH rock (ring 1) + NH def
–	1424 m	1448	1448	1441	C_7H_2 bend
1452 vs	1454 m	1451	1455	1456	C_1N stretch + CH rock (ring 1)
1470 ms	1470 m	1468	1468	1471	$C_{1'}N$ stretch + CH rock (ring 2)
1498 vs	1481 sh	1473	1482	1481	CH rock (ring 1 + 2) + NH def
1505 vs	1501 vw	1526	1524	1531	
1552 s	1553 sh	1565	1558	1551	Ring 1 stretch
1572 vs	1578 vs	1596	1572	1575	$O_1C_8O_2$ as. stretch
1590 s	1585 s	1605	1588	1599	Ring 1 + 2 stretch
1603 w	1605 s	1623	1615	1612	
–	2829 m	2887	2850	2858	CH stretch (C_7)
2920 w	2900 m	2936	2918	2949	CH stretch (ring 2)
–	2929 m	3003	2950	3009	
2967 sh	2961 m	3012	2987	3061	CH stretch (ring 1)
2980 m	2973 m	3026	3002	3084	
3040 m	3054 s	3080	3081	3092	NH stretch
3085 m	3069 s	3098	3092	3107	

Abbreviations: [a] Calculated with RHF/6-31G*, [b] Calculated with BPW91/6-31G*, [c] Calculated with B3LYP/6-31G*, ring 1 = phenyl ring with Cl atoms, ring 2 = phenyl ring, w = weak, m = medium, s = strong, v = very, sh = shoulder, def = deformation, stretch = stretching, rock = rocking, wag = wagging.

Reprinted from Chem. Phys. 298, Iliescu T, Baia M, Kiefer W, FT-Raman, surface-enhanced Raman spectroscopy and theoretical investigations of diclofenac sodium, 167–174, copyright 2004 with permission from Elsevier

The development of DFT has provided an alternative means of including electron correlation in the study of the vibrational wavenumbers of moderately large molecules (Hutter et al. 1994, Barone et al. 1995). The DFT hybrid B3LYP functional tends also to overestimate the fundamental modes in comparison to the BPW91 method, therefore scaling factors have to be used for obtaining a considerable better agreement with the experimental data (Scott and Radom 1996, Wong 1996). Thus, according to the work of Rauhut and Pulay (Rauhut and Pulay 1995), a scaling factor of 0.963 has been uniformly applied to the B3LYP calculated wavenumber values from Table 4.2. The observed disagreement between the theory and experiment could be a consequence of the anharmonicity and of the general tendency of the quantum chemical methods to overestimate the force constants at the exact equilibrium geometry (Scott and Radom 1996). Nevertheless, as one can see from Table 4.2, the theoretical calculations reproduce well the experimental data and allow the assignment of the vibrational modes. The calculated Raman intensities presented in Fig. 4.4 are also in good agreement with the experimental data.

As one can see from Fig. 4.4, the dominant bands of the FT-Raman spectrum of polycrystalline DCFNa appear at 1605 (calc. $1615\,cm^{-1}$), 1585 (calc. $1588\,cm^{-1}$), and $1578\,cm^{-1}$ (calc. $1572\,cm^{-1}$) and are given by phenyl ring stretching vibrations and the asymmetric OCO stretching modes, respectively. The ring breathing vibrations determine also intense bands at 1073 (calc. $1081\,cm^{-1}$) and $1046\,cm^{-1}$ (calc. $1051\,cm^{-1}$) (see Table 4.2). The in-plane deformation vibrations of the CH groups of both rings give rise to Raman bands at 1160 (calc. $1167\,cm^{-1}$) and $1150\,cm^{-1}$ (calc. $1149\,cm^{-1}$) (bending vibrations) and 1281 (calc. $1284\,cm^{-1}$), 1250 (calc. $1269\,cm^{-1}$) and $1235\,cm^{-1}$ (calc. $1243\,cm^{-1}$) (rocking vibrations). The medium intense Raman bands at 517 (calc. $524\,cm^{-1}$) and $533\,cm^{-1}$ (calc. $537\,cm^{-1}$) are determined by the out-of-plane deformation vibrations of the phenyl rings. The bands attributed to the out-of-plane deformation vibrations of the CH groups occur in the $840–950\,cm^{-1}$ spectral range of the Raman spectrum. In the high wavenumber region between 3069 and $2890\,cm^{-1}$ six bands assigned to the NH and CH stretching modes are observed. Weak bands at 1398 (calc. $1385\,cm^{-1}$) and $637\,cm^{-1}$ (calc. $630\,cm^{-1}$) assigned to the symmetric stretching and in-plane deformation vibration of the carboxylate group can be also seen in the Raman spectrum. The absence of the carbonyl stretching band in the $1800–1600\,cm^{-1}$ spectral range confirms the presence of the carboxylate group in the DCFNa species in the solid state (Iliescu et al. 2004a, Iliescu et al. 2003).

4.1.2 Adsorption on the Silver Surface

SERS spectra of diclofenac on the silver colloid at different pH values, together with the Raman spectrum of the polycrystalline sample, are presented in Fig. 4.5. The assignment of the vibrational modes of DCFNa to the SERS bands at different pH values is summarized in Table 4.3.

Having in view that for recording the SERS spectra, a DCFNa ethanol solution has been employed, and that in the solution the sodium atom is dissociated, we will further discuss the adsorption behavior of the diclofenac anion, which will be denoted as DCF.

As one can see from Fig. 4.5, good SERS spectra were obtained in acidic and neutral medium, while at alkaline pHs the spectra presented very broad bands. The shift in the peak position and the change in the relative intensities of the SERS bands with respect to the corresponding Raman bands, indicates a chemisorption process on the silver surface (Iliescu et al. 2004a, Iliescu et al. 2003).

By looking at the SERS spectra one can notice that the C=O stretching mode is absent in all spectra. The lack of this band evidences the presence of the carboxylate group not only in the solid state but also in the DCF adsorbed state.

Arancibia and Escadar (Arancibia and Escadar 1999) have determined from potentiometric and spectrophotometric measurements the pK_a value for the carboxylic group in DCF ($pK_a = 4.9$). Another deprotonation reaction of DCF was detected at a very acidic pH, with a pK_a value of 1.7. This value was attributed to the deprotonation of RNH_2^+ probably present in a strong acidic solution.

Taking into account the pK_a value of 4.9, an excess of DCF molecules with a carboxylic group (protonated form) is expected to be present at pH = 2. As one

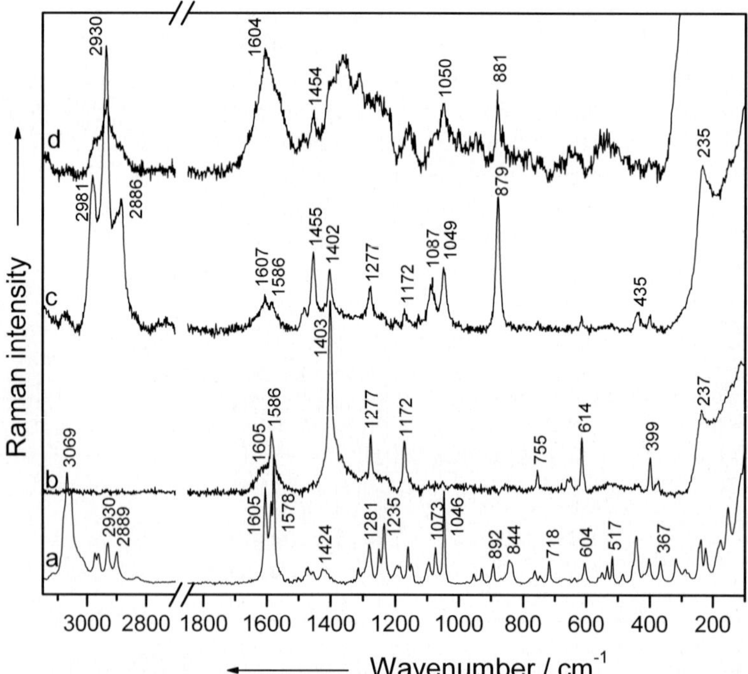

Fig. 4.5 FT-Raman spectrum of polycrystalline DCFNa (*a*) and SERS spectra of DCF on silver colloid at the pH values of 2 (*b*), 6 (*c*), and 10 (*d*). Reprinted from Chem. Phys. 298, Iliescu T, Baia M, Kiefer W, FT-Raman, surface-enhanced Raman spectroscopy and theoretical investigations of diclofenac sodium, 167–174, copyright 2004 with permission from Elsevier

can see from Fig. 4.5*b* the C=O stretching band typical for the carboxylic group is missing in the SERS spectrum at this pH value. The absence of this band could be due to the lowering of the pK_a value at the silver surface, and indicates the existence of a direct carboxylate-surface interaction. The high intensity of the symmetric and asymmetric COO^- stretching bands present in the SERS spectrum at 1403 and $1586\,cm^{-1}$ is a proof of the existence of the carboxylate group in the DCF adsorbed state and of its proximity to the silver surface. In the SERS spectrum at $pH = 2$ weak bands are also present at 647 and $399\,cm^{-1}$ that contain contributions of the in-plane COO^- deformation vibrations (see Table 4.3).

Table 4.3 Wavenumbers (in cm^{-1}) and assignment of the vibrational modes of DCFNa to the SERS bands at different pH values

Raman	SERS			Vibrational assignment
	pH 2	pH 6	pH 10	
–	237 s	235 s	–	$AgCl + AgO$ stretch
367 m	373 w	–	–	$C_1NC_{1'}$ def + $C_{6',7,8}$ def + CCl def
402 m	399 m	399 m	–	$O_1C_8O_2$ def + $Cl_1C_{2,3}$, $Cl_2C_{6,5}$ def + C_7H_2 def
442 ms	–	435 w	–	Ring 2 out-of-plane def
533 mw	–	–	537 br	Ring 2 + 1 out-of-plane def
604 m	614 m	616 w	–	Ring 1 + 2 in-plane def
637 w	647 w	–	649 wbr	$O_1C_8O_2$ def + C_7H_2 def
653 w	651 w	–	649 wbr	$C_1NC_{1'}$ def + ring 1 out-of-plane def
747 vw	755 w	754 vw	–	CH wag (ring 2 + 1)
868 w	–	879 vs	882 s	CH twist (ring 2 + 1)
1046 s	–	1049 s	1050 s	Ring 2 breathing
1073 ms	–	1087 ms	–	Ring 1 breathing
1160 ms	1172 m	1172 w	1155 mbr	CH bend (ring 1 + 2)
1281 ms	1277 m	1277 m	–	CH rock (ring 1 + 2) + C_7H_2 wag
1327 vw	–	–	1316 m	Ring 1 stretch + C_7H_2 wag
1398 vw	1403 vs	1402 m	1392 br	$C_{7,8}$ stretch + $O_1C_8O_2$ s. stretch
1454 m	–	1455 s	1454 mw	C_1N stretch + CH rock (ring 1)
1470 sh	–	1480 w	1484 w	$C_{1'}N$ stretch + CH rock (ring 2)
1578 vs	1586 ms	1586 m	–	$O_1C_8O_2$ as. stretch
1605 vs	1606 sh	1607 m	1604 sbr	Ring 1 + 2 stretch
2890 mw	–	2886 s	2888 sh	CH stretch (C_7)
2930 m	–	2934 vs	2931 s	CH stretch (ring 2)
2973 w	–	2981	2970 sh	CH stretch (ring 1)
3069 s	–	–	3062 wbr	NH stretch

Abbreviations: w = weak, m = medium, s = strong, v = very, sh = shoulder, br = broad,
stretch = stretching, bend = bending, def = deformation, wag = wagging, rock = rocking,
twist = twisting, ring 1 = phenyl ring with Cl atoms, ring 2 = phenyl ring.
Reprinted from Chem. Phys. 298, Iliescu T, Baia M, Kiefer W, FT-Raman, surface-enhanced Raman spectroscopy and theoretical investigations of diclofenac sodium, 167–174, copyright 2004 with permission from Elsevier

Many molecular species, which posses a carboxyl group, can be bonded on the metallic surface as carboxylate groups. Kwon et al. (Kwon et al. 1994a) concluded from SERS spectra of 4-(methylthio)-benzoic acid that the symmetric stretching band of the COO^- group appeared very distinctly with a broad bandwidth, when this molecular species was bonded to the silver surface via the π system of the carboxylate group. Its peak position in the adsorbed state was shifted to lower wavenumbers by as much as $6\,cm^{-1}$ compared to that of the free molecule. The asymmetric stretching mode of the carboxylate group was barely discernible in the SERS spectra, when adsorption occurred via the π system. Park et al. (Park et al. 1990) reported the adsorption of a 4-amino-benzoic acid on the silver surface via the π system of the carboxylate group. They observed a red shift of the symmetric stretching mode of the carboxylate group by approximately $10\,cm^{-1}$, compared to that of the free molecule and a very distinct SERS intensity. In contrast, the asymmetric stretching mode became weaker upon π coordination without a noticeable peak shift. On the other hand, a blue shift of the symmetric COO^- stretching mode upon surface adsorption was observed in the SERS spectrum of benzoic acid on silver electrode surface (Kwon et al. 1994b). This behavior was interpreted as a consequence of the coordination of the carboxylate group to the silver surface via its oxygen lone pair electrons.

By inspecting Fig. 4.5 and Table 4.3, blue shifts by 5 and $8\,cm^{-1}$ of the symmetric and asymmetric COO^- stretching bands were observed in the SERS spectrum of the DCF molecule in an acidic medium in comparison with the corresponding bands from the Raman spectrum, which confirm (Iliescu et al. 2004a, Iliescu et al. 2003) the binding of this molecular species on the silver surface via oxygen lone pair electrons of the carboxylate group. According to the surface selection rules for Raman scattering (Creighton 1983, Moskovits and DiLella 1980), the vibration of the adsorbed molecules, which has a polarizability tensor component normal to the surface, will be preferentially enhanced. Stretching vibrations are assumed to have a large component of the polarizability along the bond axis. The very high intensity of the symmetric and asymmetric stretching bands of the COO^- group observed in the SERS spectrum of DCF at $pH = 2$ indicates the perpendicular or at least tilted orientation of this group with respect to the silver surface.

It is known (Iliescu et al. 1995) that molecules with a nitrogen ring atom can form a pair with the chloride ion and this pair is bonded to the silver surface. By looking at the SERS spectrum recorded at a pH value of 2 one can infer that the carboxylate group of the DCF is directly bound to the silver surface, otherwise a strong change in the peak position of the AgCl stretching mode would occur.

The vibrations specific to phenyl rings are also present in the SERS spectrum in an acidic medium. The stretching vibration of both rings gives rise to a broad shoulder at $1606\,cm^{-1}$. The in-plane CH and rings deformation vibrations were observed in the SERS spectrum at 1277, 1172, and $614\,cm^{-1}$ (see Table 4.3). The shifts of these bands compared to their corresponding Raman bands confirm the interaction between phenyl rings and the silver surface. The out-of-plane deformation vibrations of CH groups of both rings present in the Raman spectrum in the spectral range between 850 and $950\,cm^{-1}$ are not present in the SERS spectrum at the pH value of 2. If one closely examines the conformation of the DCF molecule (Fig. 4.2) and the enhancement of the in-plane vibrations of phenyl rings, one can assume a tilted close to flat orientation of these rings with respect to the silver surface (Fig. 4.6a) (Iliescu et al. 2004a). According to the surface selection rules (Creighton 1983, Moskovits and DiLella 1980) one would expect the CH ring stretching modes to be present in the adsorbed state of DCF molecules with weak intensity. The absence of these bands in the SERS spectrum can be explained by the marginal contribution of these modes to the α_{zz} polarizability component (z being the axis perpendicular to the surface). A similar situation was found for the adsorbed phthalazine (Moskovits and Suh 1984), where CH stretching modes are very weak, even though the molecule stands up on the surface. A deformation of the DCF molecule in an adsorbed state could also occur.

SERS spectra recorded at close to neutral and alkaline pH values (Fig. 4.5c and d) show new bands in comparison to the spectrum obtained in an acidic environment. Very intense bands are developed in the high wavenumber region around $2900\,\mathrm{cm}^{-1}$. The peak at $1607\,\mathrm{cm}^{-1}$ that appears as a shoulder in the spectrum recorded at the pH value of 2, became, in the SERS spectrum at pH = 6, even more intense than the band at $1586\,\mathrm{cm}^{-1}$. New peaks are also developed at 1480, 1455, 1087, 1049, 879, and $435\,\mathrm{cm}^{-1}$. These bands arise also in the SERS spectrum at the pH value of 10 but with a broader shape, probably determined by different adsorption sites in an alkaline medium. By considering the pK_a value for the carboxyl group ($pK_a = 4.9$), one supposes that the carboxylate form is present in both neutral and alkaline environments, therefore it will be further analyzed only the SERS spectrum at the pH value of 6.

The appearance of new bands in the SERS spectrum at pH = 6 indicates the change in orientation of the adsorbed DCF molecule with respect to the silver surface. The bands attributed to the symmetric and asymmetric stretching modes

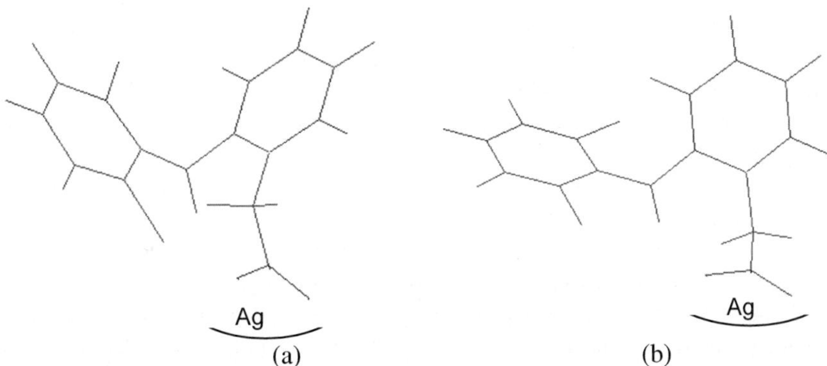

Fig. 4.6. Schematic model for the adsorption geometry of diclofenac on a colloidal silver surface at pH values (**a**) below 6 and (**b**) pH values above 6. Reprinted from Chem. Phys. 298, Iliescu T, Baia M, Kiefer W, FT-Raman, surface-enhanced Raman spectroscopy and theoretical investigations of diclofenac sodium, 167–174, copyright 2004 with permission from Elsevier

of the carboxylate group appear at 1586 and 1402 cm^{-1} in the SERS spectrum at this pH value, but with lower intensities compared to those observed in the SERS spectrum at pH = 2 (see Figs. 4.5*b* and *c*). This feature of the above-mentioned bands is a proof that the interaction between the carboxylate group and the silver surface via lone pair electrons of the oxygen atom exists also in neutral and alkaline environments, but it is weaker than in an acidic medium (Iliescu et al. 2004a, Iliescu et al. 2003).

The bands at 1049 and 1087 cm^{-1} due to the ring breathing vibrations are intense and broadened in the SERS spectrum at pH = 6 compared to the Raman spectrum. The band assigned to the C_1N stretching vibration appears at almost the same position as in the Raman spectrum (1454 cm^{-1}), but its intensity is substantially increased. The behavior of this band can be considered as an evidence of the large distance between this group and the metal surface. The C'_1N stretching band is also observed in the SERS spectrum at 1480 cm^{-1}, but with a low intensity. The presence in the SERS spectrum of the very intense bands specific to the CH ring stretching modes (around 2950 cm^{-1}), and the breathing (1049 and 1087 cm^{-1}) and stretching (1607 cm^{-1}) vibrations of the phenyl rings shows that a more perpendicular orientation of phenyl rings with respect to the silver surface exists at the pH value of 6, as compared with that taken from the adsorbed molecule in an acidic medium (see Fig. 4.6b) (Iliescu et al. 2004a).

4.1.3 Conclusions

Ab initio and DFT calculations have been performed at the RHF/6-31G*, BPW91/6-31G*, and B3LYP/6-31G* levels of theory on the two most probable conformers of DCFNa. The conformer with the sodium atom closer to the phenyl ring substituted by chlorine atoms was found to be energetically more stable by an energy difference of approximately 18 kJ/mol (RHF), 30 kJ/mol (BPW91), and 27 kJ/mol (B3LYP), respectively. Optimized structural parameters of the DCFNa calculated by various theoretical methods agree well with experimental X-ray diffraction values. The assignment of the vibrational modes was also accomplished, and a good agreement was obtained between the theoretical vibrational wavenumbers and the experimental FT-Raman data.

Good SERS spectra were obtained in acidic and neutral environments, indicating the chemisorption of the DCF molecule on the silver surface. At all studied pH values, the DCF molecule was bonded to the silver surface through the lone pair oxygen electrons of the carboxylate group, which has a perpendicular or slightly tilted orientation with respect to the silver surface. By analyzing SERS spectra at different pH values, a change of the phenyl rings' orientation with respect to the metal surface from a tilted close to flat to a more perpendicular one was concluded.

4.2 Diclofenac Sodium – β-Cyclodextrin Complex

The elucidation of the adsorption behavior of the free diclofenac molecule can be considered the starting point in the characterization of the adsorption of the DCFNa-βCD complex on the silver surface. Having in view that different parts of the DCFNa molecule can be included into the βCD cavity, the results of the investigations carried out with the SERS technique could provide insights about the complexation way.

Thus, the purpose of this work was to examine the interaction between the DCFNa and βCD in the solid state complex by using Raman spectroscopy (reprinted from Eur. J. Pharma. Sci. 22, Iliescu T, Baia M, Miclaus V, A Raman spectroscopic study of the diclofenac sodium-β-cyclodextrin interaction, 487–495, copyright 2004 with permission from Elsevier). The support of this study was the existence of some spectral ranges, where the Raman bands associated to atom group vibrations directly involved in the interaction are not overlapped. Having in mind that previous studies reported about the possibility to detect inclusional complexation by cyclodextrins at the surface of a metal (Maeda and Kitano 1995, Hill et al. 1999) SERS spectra of the DCFNa-βCD inclusion complex were also recorded and analyzed in an attempt to elucidate the adsorption behavior of the guest-host complex on the silver surface, and thus to discriminate between the possible ways of complexation. The analysis of the SERS spectra of the guest-host complex was based on the results obtained for the free DCFNa molecule.

4.2.1 Vibrational Analysis

Figure 4.7 shows the Raman spectra of the βCD, DCFNa, their guest-host complex and the 1 : 1 DCFNa-βCD physical mixture in the spectral range between 200 and 3300 cm^{-1} and the assignment of the main Raman bands of the guest molecule is summarized in Table 4.4.

By comparing the spectra illustrated in this figure one can see that the Raman spectrum of the 1 : 1 physical mixture closely resembles the sum of the individual spectra of the guest and host molecules. Furthermore, one can notice the absence of the Raman bands given by the βCD molecule vibrations in the following spectral ranges: 3050–3150, 1500–1650, 1220–1300, and 1000–1100 cm^{-1}. These spectral regions will be analyzed in detail in order to evidence the changes caused by the guest-host interaction concerning the positions and widths of the DCFNa Raman bands (Iliescu et al. 2004b, Iliescu et al. 2004c).

The Raman spectra of the βCD, DCFNa, the inclusion complex, and the 1 : 1 DCFNa-βCD physical mixture in the spectral range between 1500 and 1650 cm^{-1} are illustrated in Fig. 4.8. In this spectral region, three bands can be seen in the Raman spectrum of the DCFNa molecule.

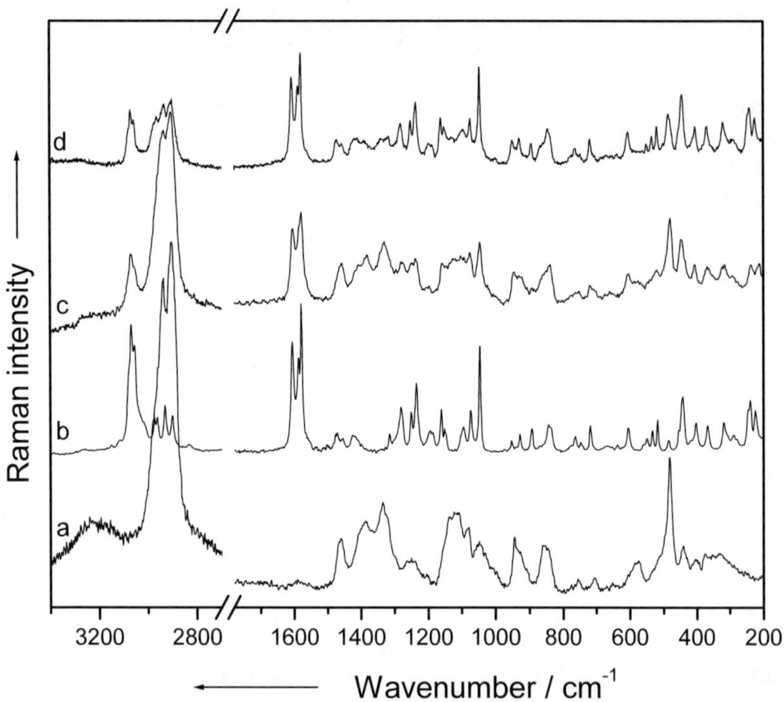

Fig. 4.7 FT-Raman spectra of βCD (*a*), DCFNa (*b*), 1 : 1 DCFNa-βCD complex (*c*), 1 : 1 DCFNa-βCD physical mixture (*d*). Reprinted from Eur. J. Pharma. Sci. 22, Iliescu T, Baia M, Miclaus V, A Raman spectroscopic study of the diclofenac sodium-β-cyclodextrin interaction, 487–495, copyright 2004 with permission from Elsevier

Thus, the band at $1578\,cm^{-1}$ was assigned to the $O_1C_8O_2$ asymmetric stretching vibration (see Fig. 4.1), while the bands at 1585 and $1605\,cm^{-1}$ were attributed to dichlorophenyl and phenylacetate rings stretching vibrations (see Table 4.4). As can be observed, the corresponding Raman bands of the inclusion complex are broader and their peak positions are changed in comparison to the bands of the pure DCFNa and indicate the existence of the guest-host interactions (Iliescu et al. 2004b, Iliescu et al. 2004c). One should emphasize the similitude of these Raman bands in the spectra of the pure DCFNa and 1 : 1 DCFNa-βCD physical mixture.

In order to separate the unresolved bands present in this spectral region into several components, the curve-fitting technique was used. In order to fit as realistically as possible, the component Raman bands were approximated by Lorentz functions, and a minimum number of bands corresponding to the number of distinct features observed in the experimental spectrum such as resolved maxima and well-developed shoulders was used. The deconvolutions of the bands corresponding to the pure DCFNa molecule and the DCFNa-βCD complex are shown in Figs. 4.9a and b, and the deconvolution data are given in the insert of each figure.

Table 4.4 Selected FT-Raman and SERS bands of DCFNa-βCD complex with their vibrational assignment. The Raman bands of the pure DCFNa are also presented

FT-Raman			SERS			Vibrational
DCFNa	DCFNa-βCD	pH 2	pH 6	pH 10		assignment[1]
–	–	216 sh	216 sh	218 vs		AgO stretch +
223 m	220 m	221 sh	222 sh	223 sh		ring 2 out-of-plane def
239 m	236 m	233 sh	233 ms	233 sh		CONa def + C_7H_2 rock +
–	–	246 s	240 sh	240 sh		AgCl stretch
–	–	381 m	–	–		$O_1C_8O_2$ def + $Cl_1C_{2,\,3}$, $Cl_2C_{6,\,5}$
402 m	404 w	407 m	–	–		def + C_7H_2 def
637 w	–	622 m	–	–		$O_1C_8O_2$ def + C_7H_2 def
653 w	658 vw	657 w	668 w	673 w		$C_1NC_{1'}$ def + ring 1 out-of-plane def
844 m	838 ms	–	886 s	883 s		CH twist (rings)
892 m	893 w	–	–	–		
1046 s	1046 s	–	1058 w	1052 w		Ring 2 breathing
1073 ms	1075 ms	–	–	–		Ring 1 breathing
1094 m	1092 m	–	1093 w	1089 w		C_7H_2 wag + CH bend (rings)
1150 m	–	–	–	–		CH bend (rings)
1160 ms	1158 m	1180 m	–	–		
1281 ms	1280 m	1284 m	1286 w	1281 w		CH rock (rings) + C_7H_2 wag
1398 vw	1405 sh	1407 s	–	–		$C_{7,\,8}$ stretch + $O_1C_8O_2$ s. stretch
1454 m	1456 ms	–	1461 m	1461 m		C_1N str + CH rock (ring 1)
1578 vs	1577 s	1577 sh	–	–		$O_1C_8O_2$ as. stretch
1585 s	1584 sh	1590 m	–	–		Ring 1 stretch
1605 s	1603 s	1613 sh	1609 mw	1609 mw		Ring 2 stretch
2829 m	–	–	2897 m	2897 m		CH stretch (C_7)
2900 m	2904 vs	–	–			
2929 m	2933 s	–	2939 s	2937 s		CH stretch (ring 2)
2961 m	–	–	–	–		
2973 m	–	–	2988 m	2986 m		
3054 s	3054 sh	–	–	–		
3069 s	3070 m	–	–	–		CH stretch (ring 1)
3080 sh	3077 sh	–	–	–		

Abbreviations: w = weak, m = medium, s = strong, v = very, sh = shoulder, stretch = stretching, def = deformation, rock = rocking, twist = twisting, wag = wagging, bend = bending, ring 1 = dichlorophenyl ring, ring 2 = phenylacetate ring, [1]Ref. (Iliescu et al. 2004a)

Reprinted from Eur. J. Pharma. Sci. 22, Iliescu T, Baia M, Miclaus V, A Raman spectroscopic study of the diclofenac sodium-β-cyclodextrin interaction, 487–495, copyright 2004 with permission from Elsevier

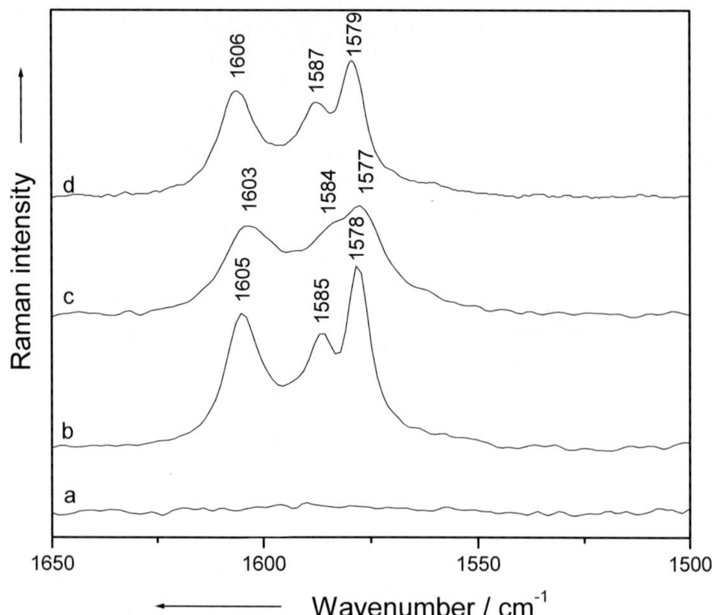

Fig. 4.8 FT-Raman spectra of βCD (*a*), DCFNa (*b*), 1 : 1 DCFNa-βCD complex (*c*), 1 : 1 DCFNa-βCD physical mixture (*d*) in the 1500–1650 cm^{-1} spectral range. Reprinted from Eur. J. Pharma. Sci. 22, Iliescu T, Baia M, Miclaus V, A Raman spectroscopic study of the diclofenac sodium-β-cyclodextrin interaction, 487–495, copyright 2004 with permission from Elsevier

The bands located at 1577.6 and 1604.8 cm^{-1} in the Raman spectrum of pure DCFNa show only a small change in their peak position after encapsulation and appear at 1576.4 and 1603.5 cm^{-1}. A broadening of these bands is also observed in the Raman spectrum of the complex, their widths being changed from 6.26 and 9.60 cm^{-1} (pure DCFNa) to 8.33 and 12.51 cm^{-1} (DCFNa-βCD complex), respectively. This behavior indicates the decrease of the vibrational relaxation time of these vibrations, in which are implied the asymmetric stretching vibration of the COO$^-$ group and the stretching vibration of the phenylacetate ring (the contribution of rotational relaxation time to the widths of these bands can be neglected). The increase of the bandwidths means the decrease of the vibrational relaxation time, and confirms the weak interaction between both the COO$^-$ group and the phenylacetate ring of the DCFNa molecule with the βCD. A more dramatic change is observed for the 1587 cm^{-1} band (FWHM 12.1 cm^{-1}), assigned to the dichlorophenyl ring stretching vibration in the spectrum of pure DCFNa. Its peak position is shifted to 1582 cm^{-1} in the spectrum of the complex and its width is changed from 12.1 cm^{-1} to 24.0 cm^{-1}.

The differences evidenced in the Raman spectra concerning the bands assigned to both rings suggest the existence of an interaction between both rings and the βCD molecule and, consequently, the existence of two isomeric 1 : 1 DCFNa–βCD

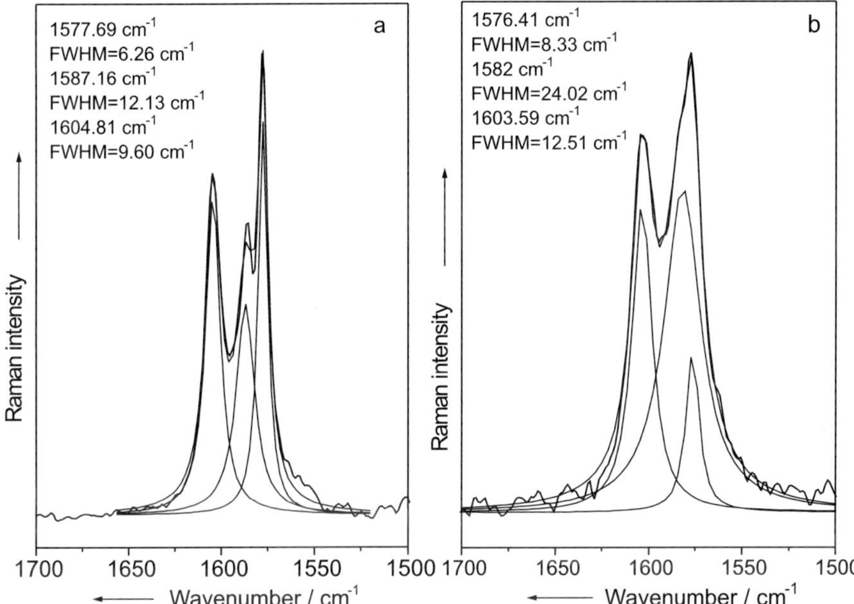

Fig. 4.9 The fit of Raman bands around (**a**) $1600\,cm^{-1}$ of DCFNa and (**b**) $1:1$ DCFNa-βCD complex. In the inset the characteristics of the bands are presented. Reprinted from Eur. J. Pharma. Sci. 22, Iliescu T, Baia M, Miclaus V, A Raman spectroscopic study of the diclofenac sodium-β-cyclodextrin interaction, 487–495, copyright 2004 with permission from Elsevier

complexes. The possibility of simultaneous inclusion of both rings of a DCFNa molecule into the βCD cavity is energetically excluded in the solid state. However, having in mind previous findings regarding the crystal structure of the inclusion complex of the sodium salt of piroxicam with βCD, which show that each host molecule takes up simultaneously two guest molecules including on the secondary hydroxyl end the benzene ring and on the primary hydroxyl end the pyridine ring of two adjacent piroxicam anions (Chiesi-Villa et al. 1998), the concomitant encapsulation of both dichlorophenyl and phenylacetate rings of two DCFNa molecules into the βCD cavity should be also considered, as much as it is possible in a solution along with the two isomeric forms of the $1:1$ complex (Mucci et al. 1999).

By comparing the changes between the bands attributed to the dichlorophenyl and phenylacetate rings vibrations, it can be supposed that the interaction of the first ring with the βCD molecule is stronger than that of the phenylacetate group with the host molecule. From infrared studies (Bratu et al. 1998) on the $1:1$ complex, it was reported that the phenylacetate ring is preferentially included in the βCD molecule due to the bulkiness of the dichlorophenyl substituent, but the changes of the bands shape of the complex were not as significant as those evidenced in the Raman spectra. Having in view the small shifts of the complex

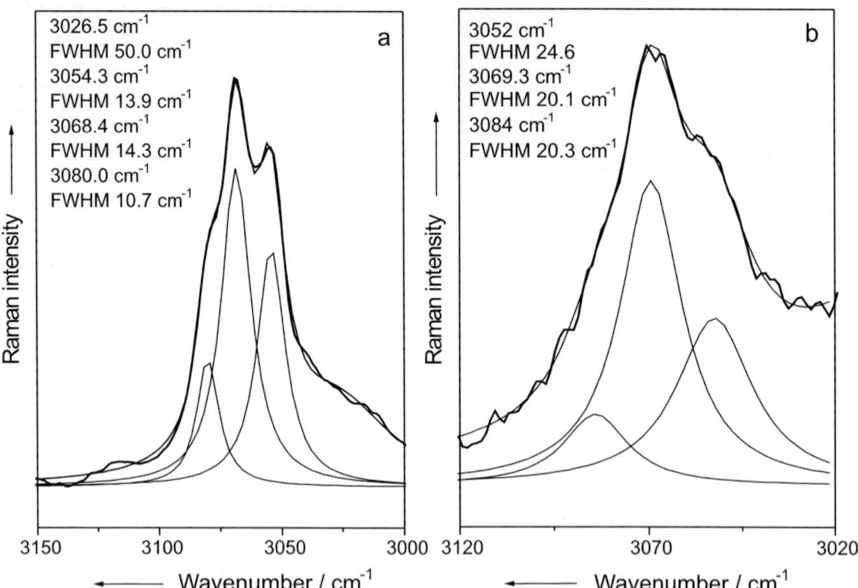

Fig. 4.10 The fit of Raman bands around (**a**) $3100\,\mathrm{cm^{-1}}$ of DCFNa and (**b**) $1:1$ DCFNa-βCD complex. In the insert the characteristics of the bands are presented. Reprinted from Eur. J. Pharma. Sci. 22, Iliescu T, Baia M, Miclaus V, A Raman spectroscopic study of the diclofenac sodium-β-cyclodextrin interaction, 487–495, copyright 2004 with permission from Elsevier

bands relative to those of the free guest molecule, the existence of hydrogen bond interactions is less probable (Bratu et al. 1998).

Further information concerning the interaction between the DCFNa rings and the host molecule is provided by the behavior of the bands attributed to the CH stretching vibrations of those rings. These bands are illustrated in Fig. 4.10 and their detailed assignment is shown in Table 4.4.

The curve-fitting technique was also used in order to separate the contributions of the individual vibrations. As can be observed, the encapsulation process induces changes of the CH stretching vibrations of the groups belonging to both rings, and supports the previous assumption of the existence of an interaction between the rings and the host molecule. By looking at the deconvolution data, one can see that the bands ascribed to the vibrations of the CH groups belonging to the dichlorophenyl ring present stronger shifts and are broader than those of the phenylacetate group. This behavior confirms (Iliescu et al. 2004b) the existence of a stronger interaction between the dichlorophenyl ring and the βCD molecule relative to that of the phenylacetate group with the host molecule in the DCFNa–βCD solid state complex.

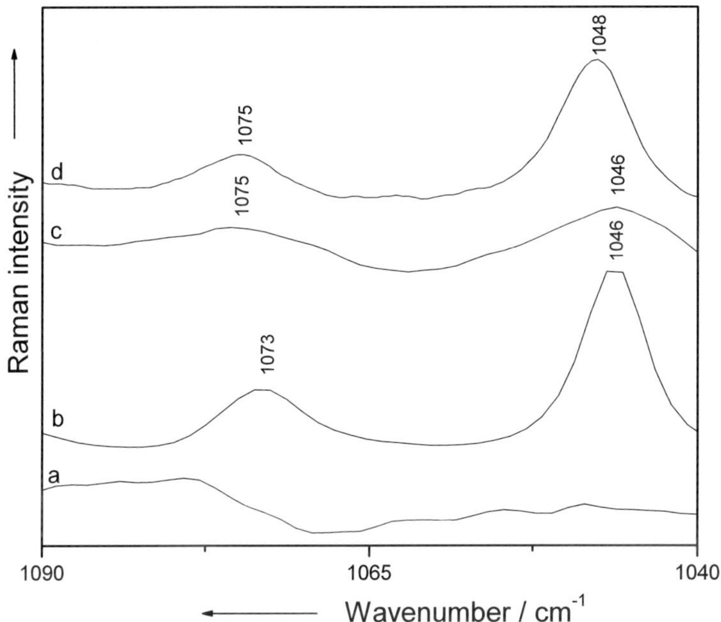

Fig. 4.11 FT-Raman spectra of βCD (*a*), DCFNa (*b*), 1:1 DCFNa-βCD complex (c), 1:1 DCFNa-βCD physical mixture (*d*) in the 1000–1100 cm^{-1} spectral range. Reprinted from Eur. J. Pharma. Sci. 22, Iliescu T, Baia M, Miclaus V, A Raman spectroscopic study of the diclofenac sodium-β-cyclodextrin interaction, 487–495, copyright 2004 with permission from Elsevier

Two other spectral ranges free of cyclodextrin Raman bands, in which are also located vibrational modes of the diclorophenyl ring and phenylacetate group, are shown in Figs. 4.11 and 4.12.

Thus, the bands from the 1000–1100 cm^{-1} spectral range, attributed to the ring breathing modes of dichlorophenyl (1073 cm^{-1}) and phenylacetate rings (1046 cm^{-1}) are presented in Fig. 4.11, while the bands from the 1220–1300 cm^{-1} spectral region (1235, 1250, and 1281 cm^{-1}) assigned to the CH rocking vibrations of both rings are illustrated in Fig. 4.12. The broadening of these bands, evidenced in the spectrum of the inclusion complex, compared with their corresponding bands from the spectra of pure DCFNa and the physical mixture reinforces the assumption of the interaction of both these molecular parts of the DCFNa with the βCD molecule in the solid-state complex form (Iliescu et al. 2004b, Iliescu et al. 2004c).

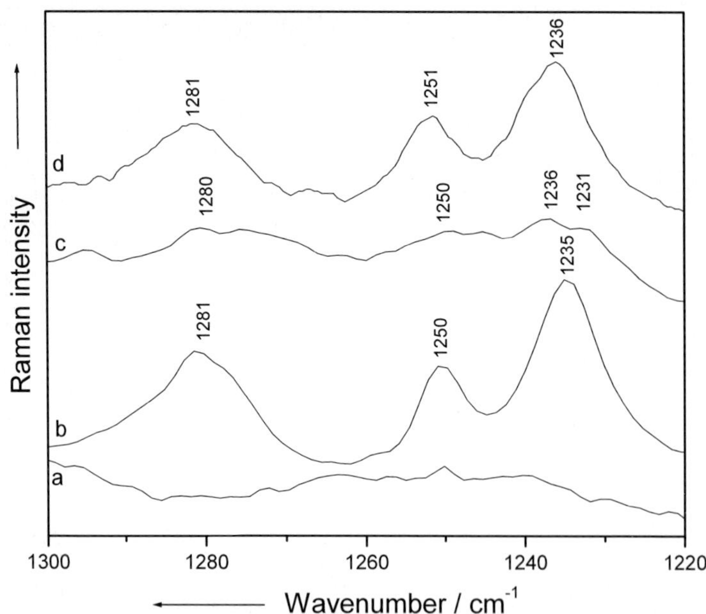

Fig. 4.12 FT-Raman spectra of βCD (*a*), DCFNa (*b*), 1 : 1 DCFNa-βCD complex (*c*), 1 : 1 DCFNa-βCD physical mixture (*d*) in the 1220–1300 cm^{-1} spectral region. Reprinted from Eur. J. Pharma. Sci. 22, Iliescu T, Baia M, Miclaus V, A Raman spectroscopic study of the diclofenac sodium-β-cyclodextrin interaction, 487–495, copyright 2004 with permission from Elsevier

4.2.2 Adsorption on the Silver Surface

By analyzing the spectral changes evidenced in the Raman spectra the detection of the inclusion complex became possible. SERS spectra should provide insights about the complexation way. Having in view that for recording the SERS spectrum a DCFNa–βCD solution was prepared and in consequence equilibrium between the free and the encapsulated guest molecules could appear, one must first verify if the recorded SERS spectrum belongs indeed to the encapsulated species. It should be mentioned that attempts to observe SERS spectra from native βCD yielded to negative results (Maeda and Kitano 1995). Therefore, the SERS spectrum of the guest–host complex was compared with the already discussed spectrum of the pure DCFNa species (Iliescu et al. 2004a) and the significant changes evidenced between the spectra demonstrate that the encapsulated guest molecule interacts with the silver surface.

The SERS spectra of the DCFNa–βCD complex in a silver colloid recorded at different pH values are shown in Fig. 4.13 along with the FT-Raman spectrum. The observed SERS bands with their assignment are presented in Table 4.4.

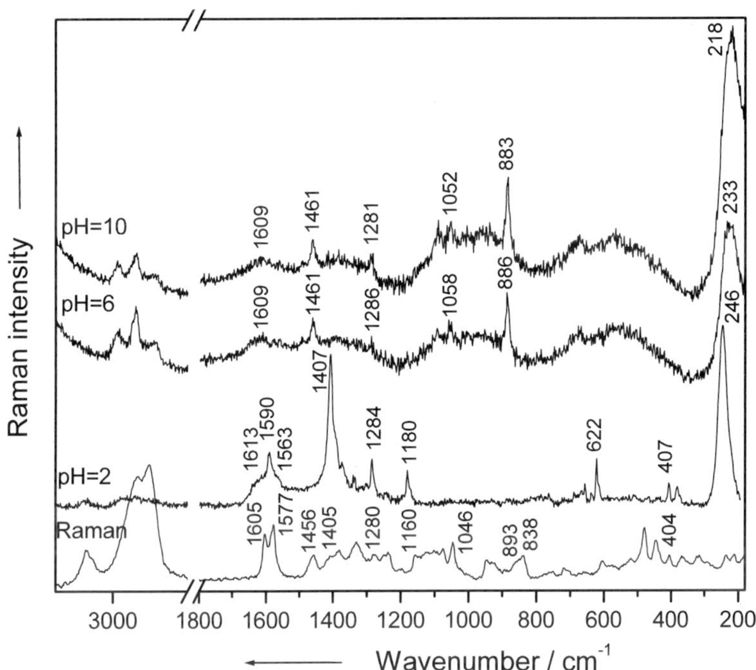

Fig. 4.13 FT-Raman and SERS spectra of the DCFNa-βCD complex at different pH values as indicated. Reprinted from Eur. J. Pharma. Sci. 22, Iliescu T, Baia M, Miclaus V, A Raman spectroscopic study of the diclofenac sodium-β-cyclodextrin interaction, 487–495, copyright 2004 with permission from Elsevier

The changes evidenced between the SERS and Raman spectra reveal the existence of a strong interaction between the encapsulated DCFNa molecule and the metal surface (Iliescu et al. 2004b, Iliescu et al. 2004c). By looking at the spectra recorded for different pH values, one can see dramatic changes on passing from an acidic to an alkaline environment. Therefore, the adsorption behavior of the guest-host complex on the colloidal silver particles will be discussed for pH values below and above 6.

From the comparison of the SERS spectrum at pH = 2 with the normal Raman spectrum, one can observe that the bands attributed to the COO$^-$ group vibrations and to the CH in-plane deformation vibrations are especially enhanced. Thus, the bands observed at 1407 and 1563 cm^{-1} in the SERS spectrum, which are due to the symmetric and asymmetric stretching vibrations of the COO$^-$ group, appear at higher wavenumbers compared to their corresponding Raman bands and are enhanced. Furthermore, the bands from 622 and 407 cm^{-1} ascribed to the deformation vibrations of the COO$^-$ group show higher intensity in the SERS spectrum relative to the Raman spectrum. The behavior of these bands indicates that the COO$^-$ group directly interacts with the silver surface.

In the low wavenumber range of the SERS spectrum, where the bands specific to the metal-molecule stretching vibration usually appear, a strong band at

$246\,\mathrm{cm}^{-1}$ can be seen. This band can be undoubtedly assigned to the AgCl stretching vibration (Sanchez-Cortez and Garcia-Ramos 1992) because in an acidic environment there is an increased amount of chloride anions caused by the addition of HCl for adjusting the pH values. However, the weak shoulder present at $218\,\mathrm{cm}^{-1}$ and assigned to the AgO stretching vibration (Chowdhury et al. 2000) together with the enhancement and the shifts of the bands attributed to the COO^- group vibrations evidenced in the SERS spectrum prove that at an acidic pH the adsorption of the guest molecule on the silver surface is maintained through the lone pair electrons of the oxygen atom (Iliescu et al. 2004b, Iliescu et al. 2004c).

In the spectral region around $1600\,\mathrm{cm}^{-1}$ one can notice the high intensity and the blue shift of the band, due to the stretching vibration of the dichlorophenyl ring, while the band given by the stretching vibration of the phenylacetate ring appears as a shoulder at $1613\,\mathrm{cm}^{-1}$.

By considering the spectral features evidenced in the SERS spectrum recorded at the pH value of 2, one can assume that the isomeric form of the DCFNa–βCD complex having the phenylacetate ring included into the βCD cavity is preferentially adsorbed on the metal surface in an acidic environment (see Fig. 4.14a). As mentioned before, the surface selection rules (Moskovits and Suh 1984, Hallmark and Campion 1986, Moskovits and DiLella 1980) predict that if the molecular z-axis is normal to the metal surface than the vibrational modes with their polarizability tensor component perpendicular on the surface will be preferentially enhanced compared with the modes having large x and y tensor components. By taking into account the surface selection rules and analyzing the behavior of the bands assigned to the COO^- vibrations from the SERS spectrum obtained at the pH value of 2, one can conclude that the adsorbed species are oriented relative to the silver surface in such a way that the COO^- group is perpendicular or at least tilted with respect to the surface. Furthermore, the enhancement of the bands at 1284 and $1180\,\mathrm{cm}^{-1}$ attributed to the rocking and bending vibrations of the CH groups demonstrates that the dichlorophenyl ring is tilted relative to the metal surface (Fig. 4.14a) (Iliescu et al. 2004b, Iliescu et al. 2004c).

By looking at the SERS spectrum recorded for the $pH = 6$ one can see that most of the bands due to the phenylacetate ring vibrations are enhanced. Thus, in the high wavenumber region ($2900-3200\,\mathrm{cm}^{-1}$) one can notice the presence of the medium intense bands attributed to the CH stretching vibrations of this ring. The band ascribed to the stretching vibration of the phenylacetate group appears also weakly enhanced in the SERS spectrum at $1609\,\mathrm{cm}^{-1}$. The breathing vibration of the phenylacetate ring gives rise to the band observed at $1058\,\mathrm{cm}^{-1}$ in the SERS spectrum, which is also enhanced and blue shifted by more than $10\,\mathrm{cm}^{-1}$, compared to its analogue band from the Raman spectrum. The new band at $1461\,\mathrm{cm}^{-1}$ observed in all spectra recorded for pH values close to neutral and alkaline is due to the CN stretching vibration.

In the low wavenumber range of the SERS spectrum at the pH value of 6, one can see the high intensity of the band at $218\,\mathrm{cm}^{-1}$, assigned to the AgO stretching vibration (Chowdhury et al. 2000), which proves that the guest molecules are adsorbed on the silver particle surface via the nonbonding electrons of the oxygen atom.

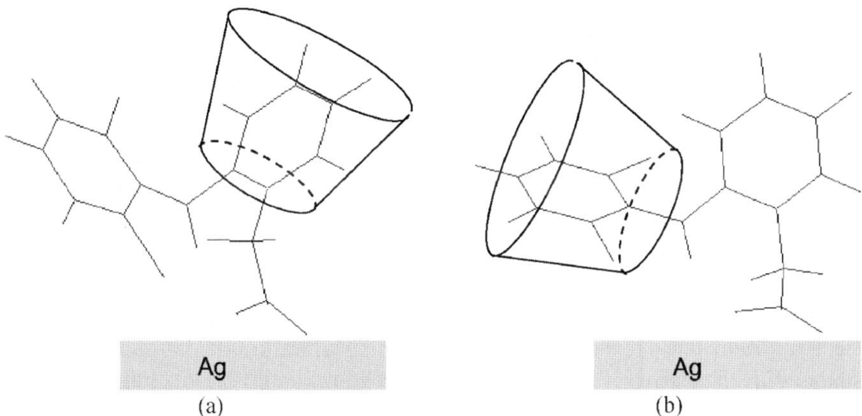

Fig. 4.14 Schematic model for the adsorption geometry of DCFNa-βCD complex on a colloidal silver surface at pH values (**a**) below 6 and (**b**) pH values above 6. Reprinted from Eur. J. Pharma. Sci. 22, Iliescu T, Baia M, Miclaus V, A Raman spectroscopic study of the diclofenac sodium-β-cyclodextrin interaction, 487–495, copyright 2004 with permission from Elsevier

All these spectral characteristics of the SERS spectra recorded for pH $\geq$ 6 indicate that at these pH values the DCFNa–βCD complex form with the dichlorophenyl ring included into the βCD cavity is mainly adsorbed on the metal surface through the nonbonding electrons of the oxygen atom (see Fig. 4.14b). According to the surface selection rules (Moskovits and Suh 1984, Hallmark and Campion 1986, Moskovits and DiLella 1980) and taking into account that the bands due to the COO^- vibrations are not enhanced in the SERS spectra one can suppose that this group is tilted or almost flat relative to the metal surface. Moreover, the enhancement of the band observed at $886\,cm^{-1}$ in the SERS spectrum, which is due to the twisting vibration of the CH groups, along with the behavior of the other bands attributed to the phenylacetate ring vibrations, suggest the tilted orientation of this ring with respect to the silver surface (Fig. 4.14b) (Iliescu et al. 2004b, Iliescu et al. 2004c).

At all pH values the bands given by the included ring vibrations are not enhanced as a consequence of the relatively large distance between these rings and the metal surface. Similar evidence for decreasing of the SERS intensity or absence of some vibrational modes attributed to vibrational groups separated several ångstroms from the metal surface has been also reported (Moskovits and Suh 1985).

4.2.3 Conclusions

The interaction between the DCFNa and βCD molecules in the solid state complex form was evidenced by means of Raman spectroscopy. Changes in the peak positions and the widths of the Raman bands of the complex compared with their corresponding bands of the pure DCFNa and physical mixture were observed. Raman data revealed the existence of interactions between both the dichlorophenyl ring and the phenylacetate group of the DCFNa species and the βCD molecule.

The changes evidenced in the SERS spectra recorded at different pH values of the solution demonstrated that, depending on the pH values, different isomeric forms of the guest-host complex are preferentially adsorbed on the silver surface. Thus, at pH values below 6 the isomeric form having the phenylacetate ring included in the βCD cavity is adsorbed on the metal surface, while at pH values equal and above 6 the isomer with the dichlorophenyl ring included into the host molecule cavity is preferentially adsorbed on the silver surface. The adsorption of the guest molecule on the metal surface is maintained in both cases through the nonbonding electrons of the oxygen atom. The probable orientation of the adsorbed species relative to the silver surface was also indicated.

References

Abdel Fattah SA, El-Khateeb SZ, Abdel Razeg SA, Tawakkol MS (1988) Application of proton magnetic resonance spectrometry in the analysis of diclofenac sodium and its tablets. Spectrosc Lett 21:533–539

Abdel-Hamid ME, Novotny L, Hamza H (2001) Determination of diclofenac sodium* flufenamic acid* indometacin and ketoprofen by LC-APCI-MS. J Pharm Biomed Anal 24:587–594

Amado AM, Moreira da Silva AM, Ribeiro-Claro PJA, Teixeira-Dias JJC (1994) Meta substituted styrene molecules included in cyclodextrin: a Raman spectroscopic study. J Raman Spectrosc 25:599–605

Amado AM, Moreira da Silva AM, Ribeiro-Claro PJA, Teixeira-Dias JJC (2000) Selection of substituted benzaldehyde conformers by the cyclodextrins inclusion process: a Raman spectroscopic study. J Raman Spectrosc 31:971–978

Arancibia JA, Escadar GM (1999) Complexation study of diclofenac with β-cyclodextrin and spectrofluorimetric determination. Analyst 124:1833–1838

Arrawal YK, Upadyay VP, Menon SK (1988) Spectrophotometric determination of diclofenac sodium. Indian J Pharma Sci 50:58–60

Astilean S, Ionescu C, Cristea Gh, Farcas SI, Bratu I, Vitoc R (1997) NMR spectroscopy of inclusion complex of sodium diclofenac with β-cyclodextrin in aqueous solution. Biospectroscopy 3:233–239

Barone V, Adamo C, Lejl F (1995) Conformational behavior of gaseous glycine by a density functional approach. J Chem Phys 102:364–370

Bratu I, Astilean S, Ionescu C, Indrea E, Huvenne JP, Legrand P (1998) FT-IR and X-ray spectroscopic investigations of diclofenac-cyclodextrins interactions. Spectrochim Acta A 54:191–196

Caira MR, Griffith VJ, Nassimbeni LR, van Oudtshoorn B (1994) Synthesis and X-ray crystal structure of β-cyclodextrin diclofenac sodium undecahydrate, a β-CD complex with unique crystal packing arrangement. J Chem Soc Chem Commun 9:1061–1062

Chiesi-Villa A, Rizzoli C, Amari G, Delcanale M, Redenti E, Ventura P (1998) The crystal structure of the inclusion complex of the sodium salt of piroxicam with β-cyclodextrin. Supramol Chem 10:111–119

Chowdhury J, Ghosh M, Misra TN (2000) pH-dependent surface enhanced Raman scattering of 8-hydroxy quinoline adsorbed on silver hydrosol. J Colloid Interface Sci 228:372–378

Creighton JA (1983) Surface Raman electromagnetic enhancement factors for molecules at the surface of small isolated metal spheres: the determination of adsorbate orientation from sers relative intensities. Surf Sci 124:209–219

Cwiertnia B, Hladon T, Stobiecki M (1999) Stability of diclofenac sodium in the inclusion complex with β-cyclodextrin in the solid state. J Pharm Pharmacol 51:1213–1218

Dastidar SG, Ganguly K, Chaudhuri K, Chakrabarty AN (2000) The anti-bacterial action of diclofenac shown by inhibition of DNA synthesis. Int J Antimicrob Agents 14:249–251

Godbillon J, Gauron S, Metayer JP (1985) High-performance liquid chromatographic determination of diclofenac and its monohydroxylated metabolites in biological fluids. J Chromatogr 338:151–159

Grandjean D, Beolor JC, Quincon MT, Savel E (1989) Automated robotic extraction and subsequent analysis of diclofenac in plasma samples. J Pharm Sci 78:247–249

Hallmark VM, Campion A (1986) Selection rules for surface Raman spectroscopy: experimental results. J Chem Phys 84:2933–2941

Hehre WJ, Radom L, Schleyer PvR, Pople JA (1986) Ab initio molecular orbital theory. Wiley, New York

Hennig B, Steup AA, Benecke R (1987) Schnellen Routinebestimmung von Diclofenac in Plasma. Pharmazie 42:861–862

Hill W, Fallourd V, Klockow D (1999) Investigation of the adsorption of gaseous aromatic compounds at surface coated with heptakis (6-thio-6-deoxy)-β-cyclodextrin by surface-enhanced Raman scattering. J Phys Chem B 103:4707–4713

Hutter J, Luthi HP, Dieterich F (1994) Structures and vibrational frequencies of the carbon molecules C2-C18 calculated by density functional theory. J Am Chem Soc 116:750–756

Iliescu T, Baia M, Kiefer W (2004a). FT-Raman, surface-enhanced Raman spectroscopy and theoretical investigations of diclofenac sodium. Chem Phys 298:167–174

Iliescu T, Baia M, Miclăuş V (2004b) A Raman spectroscopic study of the diclofenac sodium-β-cyclodextrin interaction. Eur J Pharma Sci 22:487–495

Iliescu T, Baia M, Miclăuş V, Kiefer W (2004c) A Raman spectroscopic study of the diclofenac sodium-β-cyclodextrin interaction. Proceedings of the XIXth International Conference on Raman Spectroscopy (ICORS), CSIRO Publishing, Gold Coast Queensland, 470–471

Iliescu T, Bolboaca M, Astilean S, Maniu D, Kiefer W (2003) Surface-enhanced Raman spectroscopy and theoretical investigations of diclofenac sodium. Book of Abstracts of the Third Conference of Isotopic and Molecular Processes, Cluj-Napoca, 47

Iliescu T, Vlassa M, Caragiu M, Marian I, Astilean S (1995) Raman study of 9-methylacridine adsorbed on silver sol. Vib Spectrosc 8:451–456

Kovala-Demertzi D, Mentzafos D, Terzis A (1993) Metal complexes of the anti-inflammatory drug sodium [2-[(2,6-dichlorophenyl) amino]phenyl]acetate (diclofenac sodium). Molecular and crystal structure of cadmiu diclofenac. Polyhedron 12:1361–1370.

Kwon YJ, Lee SB, Kim K, Kim MS (1994a) Raman spectroscopy of 4-(methylthio)benzoic acid adsorbed on silver surfaces. J Mol Struct 318:25–35

Kwon YJ, Soon DH, Ahn SJ, Kim S, Kim K (1994b) Vibrational spectroscopic investigation of benzoic acid adsorbed on silver. J Phys Chem 98:8481–8487

Maeda Y, Kitano H (1999) Inclusional complexation by cyclodextrins at the surface of silver as evidenced by surface-enhanced resonance Raman spectroscopy. J Phys Chem 99:487–488

Moser P, Sallmann A, Wiesenberg I (1990). Synthesis and quantitative structure-activity relationship of diclofenac analogues. J Med Chem 33:2358–2368

Mucci A, Schenetti L, Vandelli MA, Rouzi B, Formi F (1999) Evidence of the existence of 2:1 guest-host complexes between diclofenac and cyclodextrins in D_2O solutions. A ^{1}H and ^{13}C NMR study on diclofenac/β-cyclodextrin and diclofenac/2-hydroxypropil-β-cyclodextrin systems. J Chem Res 7:414–415

Maeda Y, Kitano H (1999) Inclusional complexation by cyclodextrins at the surface of silver as evidenced by surface-enhanced resonance Raman spectroscopy. J Phys Chem 99:487–488

Moskovits M, Suh JS (1984) Surface selection rules for surface enhanced Raman spectroscopy: calculations and applications to surface-enhanced Raman spectrum of phthalazine on silver. J Phys Chem 88:5526–5530

Moskovits M, DiLella DP (1980) Surface-enhanced Raman spectroscopy of benzene and benzene-$d6$ adsorbed on silver. J Chem Phys 73:6068–6075

Moskovits, M., Suh, JS (1985) Conformation of mono- and dicarboxylic acids adsorbed on silver surfaces. J Am Chem Soc 107:6826–6829

Park H, Lee SB, Kim K, Kim MS (1990) Surface-enhanced Raman scattering of p-aminobenzoic acid at silver electrode. J Phys Chem 94:7576–7580

Pose-Vilarnovo B, Santana-Penin L, Echezarreta-Lopez M, Perez-Marcos MB, Vila-Jato JL, Torres-Labandeira JJ (1999) Interaction of diclofenac sodium with β- and hydroxypropyl-β-cyclodextrin in solution. S T P Pharm Sci 9:231–236

Rauhut G, Pulay P (1995) Transferable scaling factors for density functional derived vibrational force fields. J Phys Chem 99:3093–3100

Reck G, Faust G, Dietz G (1988) Röntgenkristallographische Untersuchungen an Diclofenac-Natrium-Strukturanalyse des Diclofenac-Natrium Tetrahydrat. Pharmazie 43:771–774

Sanchez-Cortez S, Garcia-Ramos JV (1992) SERS of cytosine and its methylated derivatives on metal colloids. J Raman Spectrosc 23:61–66

Sastry CSP, Rao ARM, Prasad THV (1987) Spectrophotometric analysis of diclofenac sodium and piroxicam and their pharmaceutical preparations. Anal Lett 20:349–359

Sastry CSP, Tipirneni ASRP, Suryanarayana MV (1989) Extractive spectrophotometric determination of some anti-inflammatory agents with methylene violet. Analyst 114:513–515

Schneider W, Degen PG (1981) Simultaneous determination of diclofenac sodium and its hydroxy metabolites by capillary column gas chromatography with electron-capture detection. J Chromatogr 217:263–271

Scott AP, Radom L (1996) Harmonic vibrational frequencies: an evaluation of Hartree–Fock, Møller-Plesset, quadratic configuration interaction, density functional theory, and semiempirical scale factors. J Phys Chem 100:16502–16513

Szejtli J (1982) Cyclodextrins and drugs. In: Cyclodextrins and their inclusion complexes. Akademiai Kiado, Budapest

Tunçay M, Çaliş S, Kaş HS, Ercan MT, Peksoy I, Hincal AA (2000) Diclofenac sodium incorporated PLGA (50:50) microspheres: formulation considerations and in vitro/in vivo evaluation. Int J Pharm 195:179–188

Todd PA, Sorkin EM (1988) Diclofenac sodium. A reppraisal of pharmacodynamic and pharmaco-kinetic properties and therapeutic efficacy. Drugs 35:244–249

Whittaker DV, Penkler LJ, Glintenkamp LA, van Oudtshoorn B, Wessels PL (1996) Diclofenac-β-cyclodextrin inclusion in solution. Proton magnetic resonance and molecular modelling studies, In: Szejtli I, Szente L (eds) Proceedings of the Eight International Symposium on Cyclodextrins, Kluwer Academic Publishers, 377–380

Wong MW (1996) Vibrational frequency using density functional theory. Chem Phys Lett 256:391–399

5 Molecules with Antibacterial Properties

5.1 Potassium Benzylpenicillin

Penicillins are well-known antibacterial drugs that have been in constant clinical use after their discovery in the 1940s. Antibiotics usually exert their effect against bacteria by interfering with their basic life functions, i.e., they interfere with protein synthesis, prevent cell wall synthesis, and inhibit DNA replication (Neu 1992). Particularly, a penicillin's derivatives exert its principal antibacterial activity by binding to penicillin-binding proteins and inhibiting cell wall synthesis (Green and Wald 1996). The basis structure of penicillin consists of a five-membered sulfur-containing thiazolidine ring fused to a beta-lactam ring (Fig. 5.1).

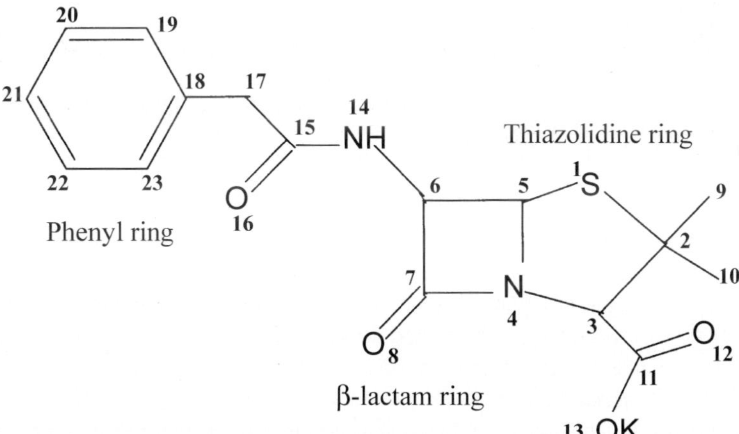

Fig. 5.1 Schematic structure of the benzylpenicillin potassium salt with the labeling of the atoms. (Raman and SERS investigations of potassium benzylpenicillin, Iliescu T, Baia M, Pavel I, copyright 2006 John Wiley & Sons Limited. Reproduced with permission)

A side chain that determines the antibacterial spectrum and other properties of the agent is attached to the beta-lactam ring. Shortly after the discovery of penicillin and its widespread use as antimicrobial agent, bacteria resistant to penicillin began to appear. Under the pressure of antibiotic usage, pathogenic bacteria, which are living organisms and yearn to survive, have to find ways to resist these antibiotics. One of the currently recognized patterns of antibacterial resistance can be explained by the general mechanisms according to which the organisms produce enzymes that destroy or inactivate antibiotics. The best-known example of such an enzyme is beta-lactamase (Coleman et al. 1994). In the presence of this enzyme, the four-membered beta-lactam ring found in penicillin is opened, preventing the compound from binding to its target, thereby rendering the antibiotic ineffective.

Owing to the increasing prevalence of antibiotic resistant bacteria, the effectiveness of a given agent may no longer be assured, leading to the need for the development of updated management strategies for patients experiencing treatment failure with antibiotics. Moreover, it was found that metals have a strong influence upon the susceptibility of bacteria to penicillin *in vitro* and, thus, monitoring the interaction between metals and antibiotics becomes of the utmost importance. Therefore, for understanding the action of drugs, it is essential to find out if the structure of the adsorbed species is similar to that of the free molecule. In these investigations, a silver surface may serve as an analogue for an artificial biological interface (Dryhurst 1977).

In the present work, DFT calculations were performed on the potassium benzylpenicillin (KBP) molecule and theoretical results were correlated with the experimental X-ray and Raman data. Raman spectra of KBP in a powder form, in a solution and adsorbed on the colloidal silver particles have been recorded and analyzed in order to get insights about the adsorption behavior on the metal surface (Raman and SERS investigations of potassium benzylpenicillin, Iliescu T, Baia M, Pavel I, copyright 2006 John Wiley & Sons Limited. Reproduced with permission). These investigations can be considered starting points for further clinical trials in order to identify optimal management strategies for specific infections.

5.1.1 Vibrational Analysis

The schematic structure of the KBP with the labeling of the atoms is illustrated in Fig. 5.1. X-ray structural investigations of the KBP molecule (Dexter and van der Veen 1978) revealed that in the solid state the side-chains of the molecule are coiled. DFT calculations have been performed on the KBP molecule at the BPW91/6–31G* theoretical level (Iliescu et al. 2006) and the selected calculated structural parameters are given in Table 5.1, together with the available X-ray data. As one can see from Table 5.1, the unscaled calculated bond lengths and bond angles agree with experimental X-ray data, the observed differences being most probably due to the intermolecular interactions, which occur in the crystal between the benzylpenicillin anion, the potassium cation, and water molecules. Moreover, the theoretical calculations were performed for the gas phase, while the experimental data were for solid phase.

Table 5.1 Selected calculated bond lengths (Å) and angles (degree) of KBP compared with the experimental data

	Calc.[a]	Exp.[b]
Bond lengths (Å)		
S_1-C_2	1.916	1.847
C_2-C_3	1.567	1.571
C_3-N_4	1.464	1.461
N_4-C_5	1.466	1.451
C_5-S_1	1.849	1.834
C_2-C_9	1.534	1.522
C_2-C_{10}	1.532	1.532
C_3-C_{11}	1.577	1.551
$C_{11}-O_{12}$	1.262	1.231
$C_{11}-O_{13}$	1.263	1.251
$O_{12}-K$	2.600	2.289
$O_{13}-K$	2.657	2.721
C_5-C_6	2.277	1.561
C_6-C_7	1.568	1.571
C_7-O_8	1.224	1.211
C_7-N_4	1.383	1.381
$C_{18}-C_{19average}$	1.401	1.351
C_6-N_{14}	1.432	1.431
$N_{14}-C_{15}$	1.377	1.341
$C_{15}-O_{16}$	1.229	1.221
$C_{15}-C_{17}$	1.535	1.512
$C_{17}-C_{18}$	1.520	1.452
Angles (degree)		
$S_1-C_2-C_3$	103.4	106.26
$C_2-C_3-N_4$	105.42	105.27
$C_3-N_4-C_5$	116.67	119.47
$N_4-C_5-S_{1'}$	105.20	105.16
$C_5-S_1-C_2$	93.82	95.24
$C_2-C_3-C_{11}$	120.85	113.97
$C_3-C_{11}-O_{12}$	119.65	117.1
$C_3-C_{11}-O_{13}$	113.22	116.1
$O_{12}-C_{11}-O_{13}$	126.92	127.1
$C_5-C_6-C_7$	98.140	84.17
$C_6-C_7-O_8$	136.011	137.1
$C_6-C_7-N_4$	91.649	93.1
$C_7-N_4-C_5$	95.077	94.18
$N_4-C_5-C_{6'}$	88.471	88.37

Table 5.1 (Continued)

	Calc.[a]	Exp.[b]
C_{18}-C_{19}-$C_{20average}$	120.0	120.18
C_5-C_6-N_{14}	117.229	117.37
C_6-N_{14}-C_{15}	122.766	121.11
N_{14}-C_{15}-C_{17}	114.60	117.1
N_{14}-C_{15}-O_{16}	122.93	121.1
C_{15}-C_{17}-C_{18}	112.124	118.1
Dihedral angles (degree)		
S_1-C_2-C_3-N_4	−36.255	−27.98
C_2-C_3-N_4-C_5	43.194	32.1
S_1-C_5-N_4-C_7	115.094	114.37
S_1-C_5-C_6-C_7	−100.953	−109.28
C_7-C_6-N_{14}-C_{15}	−117.518	−94.1
N_4-C_3-C_{11}-O_{12}	60.177	40.1
N_4-C_3-C_{11}-O_{13}	−115.034	−141.88
C_{15}-C_{17}-C_{18}-C_{23}	−115.490	−91.2

Abbreviations: [a] Calculated with BPW91/6-31 + G*, [b] Ref. (Dexter and van der Veen 1978) (Raman and SERS investigations of potassium benzylpenicillin, Iliescu T, Baia M, Pavel I, copyright 2006 John Wiley & Sons Limited. Reproduced with permission)

FT-Raman spectra of the polycrystalline KBP and its water solution at a pH value of 6 (resulted after dissolution) are illustrated in Fig. 5.2. The observed Raman bands with their assignment accomplished with the help of theoretical calculations are presented in Table 5.2.

The observed disagreement between the calculated wavenumber values and the experimental ones could be a consequence of the anharmonicity and of the general tendency of the quantum chemical methods to overestimate the force constants at the exact equilibrium geometry. Nevertheless, as can be seen from Table 5.2, the theoretical calculations reproduce the experimental data well and allow the assignment of the vibrational modes.

By inspecting Fig. 5.2, one can see that the dominant bands of the FT-Raman spectra of the polycrystalline KBP as well as its water solution appear at 1005 (calc. 1006 cm^{-1}) and 1603 cm^{-1} (calc. 1602 cm^{-1}) and are given by the stretching vibrations of the $C_{6,7}$ + $C_3N_4C_7$ bonds (for numbering of the atoms, see Fig. 5.1) and phenyl ring, respectively (Iliescu et al. 2006). Other intense bands, which are due to the asymmetric and symmetric stretching vibrations of the CH groups from beta-lactam and phenyl rings, can be observed in the spectral range between 2920 and 3080 cm^{-1}. The bands assigned to the CH_2 and CH_3 asymmetric stretching vibrations are also present with strong intensity in this wavenumbers region. The phenyl ring breathing vibration determines a medium-intense band at 1032 cm^{-1} (calc. 1027 cm^{-1}), while the trigonal ring stretching vibration gives rise to the Ra-

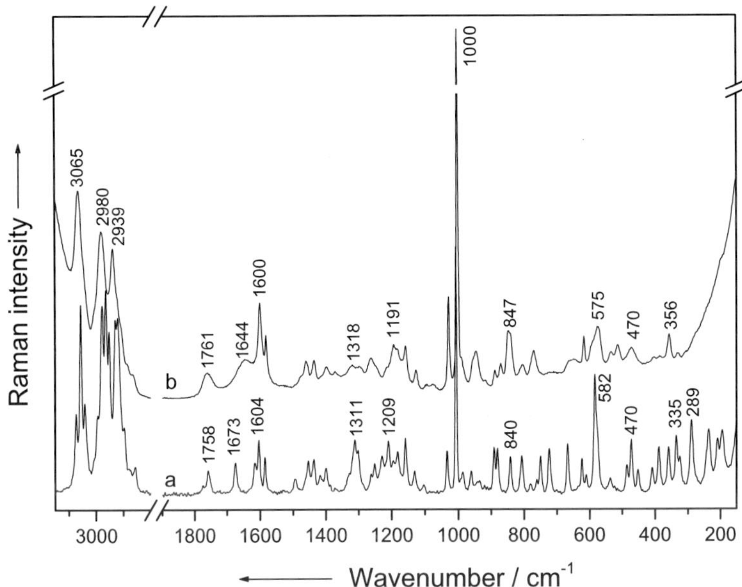

Fig. 5.2 Raman spectra of the KBP: solid state (*a*) and water solution (*b*). (Raman and SERS investigations of potassium benzylpenicillin, Iliescu T, Baia M, Pavel I, copyright 2006 John Wiley & Sons Limited. Reproduced with permission)

man band at $985\,cm^{-1}$ (calc. $981\,cm^{-1}$). The in-plane deformation vibrations of the phenyl ring together with the $C_7N_4C_3$ deformation and the out-of-plane deformation vibrations of beta-lactam and thiazolidine rings give the bands at 610 (calc. $616\,cm^{-1}$) and $623\,cm^{-1}$ (calc. $620\,cm^{-1}$). The out-of-plane deformation modes of the phenyl ring occur as medium and medium-to-weak Raman bands as 388 (calc. $396\,cm^{-1}$) and $451\,cm^{-1}$ (calc. $459\,cm^{-1}$), respectively.

The NH stretching vibration of the secondary amide group can be observed as a medium intense band in the solid-state Raman spectrum at $3371\,cm^{-1}$ (calc. $3486\,cm^{-1}$). According to Dexter and van der Veen (Dexter and van der Veen 1978) the amide group in benzylpenicillin has the trans configuration. The amide I band ($C_{15}O_{16}$ stretching mode) appears with medium intensity in the solid-state Raman spectrum at $1673\,cm^{-1}$ (calc. $1692\,cm^{-1}$). The amide II band in which the NH bending mode has the major contribution is located at $1493\,cm^{-1}$ (calc. $1492\,cm^{-1}$). Another amide band concurs with other bands given by phenyl and beta-lactam rings vibrations.

The symmetric stretching vibration of the COO^- group gives the medium-strong band at $1616\,cm^{-1}$ (calc. $1611\,cm^{-1}$) (the strongest amplitude was observed for the $C_{11}O_{13}$ bond). As expected, the band due to the asymmetric stretching vibration of this group is very weak in the Raman spectrum and concurs with CH_3 deformation bands (Iliescu et al. 2006).

Table 5.2 Assignment of the theoretical wavenumber values (cm^{-1}) to the experimental bands of the KBP molecule

Raman solid sample	Solution	Calc.	Vibrational assignment
209 m	–	215	Ring 1 out-of-plane def + OKO stretch + CH$_3$ rock
236 m	–	257	CH$_3$, CH$_2$ rock + N$_{14}$C$_{15, 16}$ twist + C$_7$O$_8$ wag
276 sh	–	284	CH$_3$ rock + S$_1$C$_{2, 10}$ wag
288 m	304 sh	298	CH$_2$ rock + C$_{17, 18, 36}$ bend
324 sh	–	320	CH$_3$ rock + S$_1$C$_{2, 9}$ twist
335 m	330 w	331	CH$_2$ rock + C$_{9,2, 10}$ bend
358 m	356 w	359	CH$_2$ rock + C$_{9,2, 3}$ twist
388 m	385 w	396	Ring 1 out-of-plane def
408 vw	402 w	411	CH$_2$ rock + N$_{14}$H wag + C$_{15}$N$_{14}$C$_6$ wag
451 w	458 sh	459	Ring 1 out-of-plane def + CH$_2$ rock
471 m	470 w	464	C$_{9,2, 10}$ bend + N$_4$C$_5$S$_1$ bend
485 w	476 sh	481	N$_{14}$H wag + ring 1 out-of-plane def
513 vw	512 m	525	Ring 3 breathing
522 vw	–	543	N$_{14}$H wag
535 w	532 w	569	CH$_2$ rock + CH def (ring 2 + 3) + C$_2$S$_1$C$_5$ stretch.
583 s	575 mw	599	C$_{15, 17}$ stretch + C$_{3, 2}$ stretch + N$_{14}$H wag
–	589 sh	–	
610 w	–	616	Ring 1 in-plane def + C$_7$N$_4$C$_3$ bend
623 mw	617 w	620	Ring 1 in-plane def + ring 2 + 3 out-of-plane def
–	648 w	–	
667 mw	660 w	656	Rings 2 + 3 out-of-plane def
691 vw	–	684	CH wag (ring 1)
700 vw	–	707	
723 m	–	721	CH wag (ring 1) + C$_5$N$_4$C$_3$, C$_{6, 5}$N$_4$ bend
731 sh	–	–	
750 m	–	752	CH wag (ring 1) + ring 2 out-of-plane def
761 wsh	770 w	758	CH wag (ring 1) + C$_{15, 17, 18}$ def + O$_{13}$C$_{11}$O$_{12}$ wag
780 w	–	788	O$_{13}$C$_{11}$O$_{12}$ bend + ring 2 in-plane def
798 sh	804 w	803	C$_{2, 3}$ stretch
805 m	–	–	
829 sh	–	819	CH twist (ring 1)
840 m	844 shm	827	CH twist (ring 1) + C$_{15, 17, 18}$ bend
–	847 ms	–	
862 vw	–	850	Ring 2 breathing + N$_4$C$_{3, 11}$ bend
880 m	870 w	885	CH twist (ring 1)
890 m	887 w	899	CH$_2$ rock + C$_{5, 6}$ stretch

Table 5.2 (Continued)

Raman solid sample	Solution	Calc.	Vibrational assignment
912 vw	–	919	$C_{3,11}$ stretch
919 vw	920 vw	–	
935 vw	945 m	965	$C_{15,17}$ stretch + $C_{6,5}$ stretch + $C_{3,11}$ stretch
960 w		–	
985 w	989 shw	991	Ring 1 trigonal stretch
1005 s	1001 vs	1006	$C_3N_4C_7$ stretch + $C_{7,6}$ stretch + CH_3 twist
1032 m	1028 ms	1027	Ring 1 breathing
1051 vw	–	1072	CH rock (ring 1) + CH_2 twist
1063 vw	1068 vw	1083	C_5N_4 stretch + CH def (rings 2 + 3)
1086 vw	–	1098	CH_3 wag + C_5N_4 stretch + $C_2S_1C_5$ stretch
1102 vw	1092 vw	1107	CH_2 twist + CH rock (ring 1) + CH def (ring 2) + $N_{14}H$ bend
1130 w	1125 m	1142	CH def (rings 2 + 3) + $C_{9,2,10}$ stretch
1157 m	1157 m	1156	CH bend (ring 1)
1181 mw	1182 shw	1182	CH def (ring 2)
1193 w	1191 m	1195	$C_{2,3}$ stretch + CH def (ring 2) + CH_3 wag
1209 m	1213 sh	1233	CH def (rings 2 + 3) + $C_3N_4C_7$ stretch
1228 mw	–	1237	$N_{14}H$ bend + $C_{15}N_{14}$ stretch
1250 w	1242 sh	1251	CH def (rings 3 + 2)
–	1246 sh	–	
1261 vw	1261 w	1271	CH_2 wag
1287 sh	–	1278	CH def (rings 2 + 3)
1301 shm	1299 w	1303	
1311 ms	1318 w	1317	$C_{3,11}$ stretch
1328 sh	–	1345	CC stretch (ring 1)
1373 m	1371 w	1349	$C_3N_4C_7$ stretch + CH def (rings 3 + 2)
1387 sh	–	1362	CH_3 wag
1399 m	1397 w	1381	
1411 sh	1415 sh	1446	
1417 w	–	1448	CH_3 bend
1436 m	1436 w	1453	
1453 m	1454 sh	1479	
–	1460 m	–	
–	1476 sh	–	
1493 m	1494 vw	1492	$N_{14}H$ bend + $C_{15}N_{14}$ stretch + CH rock (ring 1) amide II trans
1585 m	1581 ms	1584	CC stretch (ring 1)
1603 ms	1600 s	1602	

Table 5.2 (Continued)

Raman solid sample	Solution	Calc.	Vibrational assignment
1616 ms	1638 shw	1611	$C_{11}O_{13}$ stretch
–	1644 wm	–	
1673 m	–	1692	$C_{15}O_{16}$ stretch
1758 mw	1761 mw	1744	C_7O_8 stretch
1775 m	–	–	
2856 shw	2865 sh	2972	CH_2 s. stretch
2919 s	2938 s	3021	CH as. stretch (ring 2)
2928 s	–	3037	CH_2 as. stretch
2951 s	–	3042	CH_3 as. stretch
2964 s	–	3047	
2976 s	2981 s	3084	CH as. stretch (ring 1)
3041 m	–	3105	
3055 s	–	3116	
3073 s	3065 s	3123	CH s. stretch (ring 1)
3371 m	–	3486	NH stretch

Abbreviations: [a] Calculated with BPW91/6-31 + G*, v = very, w = weak, m = medium, s = strong, sh = shoulder, def = deformation, rock = rocking; bend = bending; wag = wagging, twist = twisting; stretch = stretching; ring 1 = phenyl ring, ring 2 = beta-lactam ring, ring 3 = thiazolidine ring.
(Raman and SERS investigations of potassium benzylpenicillin, Iliescu T, Baia M, Pavel I, copyright 2006 John Wiley & Sons Limited. Reproduced with permission)

The thiazolidine ring vibrations are represented in the Raman spectrum by the medium-strong band at 583 cm^{-1} (calc. 599 cm^{-1}) assigned to the C_2C_3 stretching mode and the medium intense band located at 471 cm^{-1} (calc. 476 cm^{-1}) ascribed to the $N_4C_5S_1$ deformation vibration. The band due to the out-of-plane deformation vibration of the beta-lactam ring appears at 750 cm^{-1} (calc. 752 cm^{-1}).

A close analysis of Fig. 5.2 reveals no dramatic changes in the Raman spectrum of the KBP solution, in comparison with the solid-state Raman spectrum, besides the well-known characteristics of the solution spectra concerning the broadening of the bands and the change of their position.

5.1.2 Adsorption on the Silver Surface

The SERS spectrum of KBP at a pH value of 6 compared with its solution Raman spectrum at the same pH value is presented in Fig. 5.3. SERS and Raman bands together with their assignment are summarized in Table 5.3.

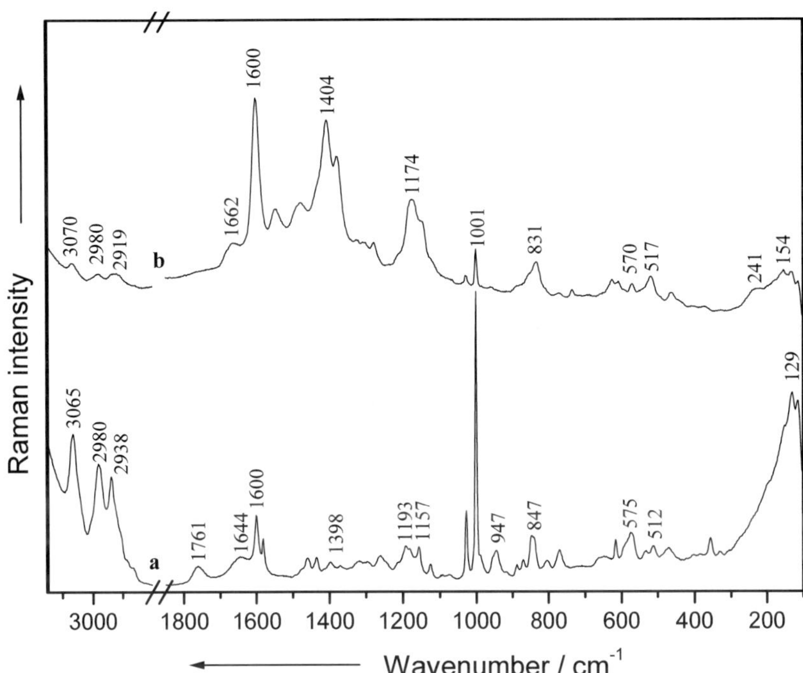

Fig. 5.3 Raman spectrum of KBP water solution (*a*) and SERS spectrum of KBP at the pH value of 6 (*b*). (Raman and SERS investigations of potassium benzylpenicillin, Iliescu T, Baia M, Pavel I, copyright 2006 John Wiley & Sons Limited. Reproduced with permission)

By inspecting Fig. 5.3, one observes dramatic changes in the relative intensities of some bands from the SERS spectrum as compared to the normal Raman spectrum. Thus, the bands at 1600, 1405, and $1174 \, \text{cm}^{-1}$ become the most intense in the SERS spectrum, while the intensity of the band located at $1001 \, \text{cm}^{-1}$ is very weak in comparison to the bulk Raman spectrum. Moreover, shifts of some SERS bands relative to the corresponding Raman bands of the KBP solution can also be seen. These spectral features confirm the chemisorption of the KBP molecule through some of its constituent groups, while other groups are located at a relatively large distance from the metal surface, and therefore their vibrations are not influenced by adsorption (Iliescu et al. 2006).

The orientation of the adsorbed KBP molecule can be deduced by using surface selection rules for Raman scattering (Moskovits and DiLella 1980, Gao and Weaver 1985). According to them, if the molecular axis (*z*-axis) is normal to the surface, then vibrations of the adsorbed molecule, which have a polarizability tensor component along this axis, will be preferentially enhanced. The KBP molecule has very low symmetry and all vibrations have a polarizability tensor component along the *z*-axis. Stretching vibrations, however, are assumed to have the largest component along the bond axis. Stretches that are orthogonal to the surface will result in different intensities in SERS and bulk Raman spectra. In addition,

Table 5.3 Wavenumbers (cm^{-1}) and assignment of the vibrational modes of KBP molecule to the SERS bands at the pH value of 6

Raman solution	SERS pH 6	Vibrational assignment
356 w	–	CH_2 rock + $C_{9,2,3}$ twist
470 w	461 mw	$C_{9,2,10}$ bend + $N_4C_5S_1$ bend
512 w	519 m	Ring 3 breathing
532 w	–	CH_2 rock + CH def (ring 2 + 3) + $C_2S_1C_5$ stretch
575 mw	570 w	$C_{15,17}$ stretch + $C_{3,2}$ stretch + $N_{14}H$ wag
617 w	625 mw	Ring 1 in-plane def + ring 2 + 3 out-of-plane def
770 w	767 w	CH wag (ring 1) + $C_{15,17,18}$ def + $O_{13}C_{11}O_{12}$ wag
804 w	–	$C_{2,3}$ stretch
847 m	831 m	CH twist (ring 1) + $C_{15,17,18}$ bend
870 w	876 sh	CH twist (ring 1)
887 w	–	CH_2 rock + $C_{5,6}$ stretch
920 vw	–	$C_{3,11}$ strech
945 m	961 sh	$C_{15,17}$ stretch + $C_{6,5}$ stretch + $C_{3,11}$ stretch
989 sh	–	Ring 1 trigonal stretch
1001 vs	1001 m	$C_3N_4C_7$ stretch + $C_{7,6}$ stretch + CH_3 twist
1028 ms	1029 w	Ring 1 breathing
1075 vw	1065 sh	C_5N_4 stretch + CH def (rings 2 + 3)
1092 vw	–	CH_2 twist + CH rock (ring 1) + CH def (ring 2) + $N_{14}H$ bend
1125 w	1118 sh	CH def (rings 2 + 3) + $C_{9,2,10}$ stretch
1157 m	1149 sh	CH bend (ring 1)
1182 sh	1174 s	CH def (ring 2)
1191 m	–	$C_{2,3}$ stretch + CH def (ring 2) + CH_3 wag
1261 w	1277 mw	CH_2 wag
1299 w	1308 w	CH def (rings 2 + 3)
1318 w	–	$C_{3,11}$ stretch
1371 w	1379 sh	$C_3N_4C_7$ stretch + CH def (rings 3 + 2)
1397 w	1405 m	CH_3 wag
1460 w	1477 m	CH_3 bend
1600 s	1600 vs	CC stretch (ring 1)
1644 mw	1664 mw	$C_{11}O_{13}$ stretch
1761 mw	–	C_7O_8 stretch
2938 s	2919 w	CH as. stretch (ring 2)
2981 s	2981 w	CH as. stretch (ring 1)
3065 s	3066 w	CH s. stretch (ring 1)

Abbreviations: v = very; w = weak; m = medium; s = strong; sh = shoulder; rock = rocking; bend = bending: wag = wagging; twist = twisting; def = deformation, stretch = stretching; ring 1 = phenyl ring, ring 2 = beta-lactam ring, ring 3 = thiazolidine ring
(Raman and SERS investigations of potassium benzylpenicillin, Iliescu T, Baia M, Pavel I, copyright 2006 John Wiley & Sons Limited. Reproduced with permission)

monosubstituted benzene ring modes were shown to be sensitive to the orientation of the ring plane with respect to the metal surface in terms of line position and shape (Gao and Weaver 1985).

The bonding of the KBP molecule to the silver surface can be realized through different atoms. Thus, it is known (Wei et al. 2001, Hu et al. 2005) that a sulfur atom can form a stable metal-sulfur bond. Moreover, the KBP molecule could be adsorbed on the Ag surface either through the nitrogen atom of the amide group and/or the oxygen atoms of the amide, carbonyl, or carboxylate group, respectively. These adsorption possibilities will be further discussed bearing in mind the predictions of surface selection rules and analyzing the intensity of the SERS bands.

The absence in the SERS spectrum of the bands due to vibrations in which the sulfur atom from the thiazolidine ring is involved (535 and $1086\,cm^{-1}$) (Table 5.2) proves that this atom is not implicated in the KBP adsorption on the silver surface. The very weak intensity of all amide SERS bands (Fig. 5.3 and Table 5.2) indicates that this group is not in the proximity of the silver surface. Moreover, the absence in the SERS spectrum of the carbonyl stretching vibrations evidenced in the solid-state Raman spectrum at 1673 ($C_{15}O_{16}$) and $1758\,cm^{-1}$ (C_7O_8) suggests that these groups are not bonded to the metal surface (Iliescu et al. 2006).

On the other hand, the presence of the medium-intense band located at $1664\,cm^{-1}$ in the SERS spectrum and assigned to the stretching vibration of the carboxylate group is an indication of the possible involvement of this group in the KBP adsorption. The shift of this band from 1644 to $1664\,cm^{-1}$ reinforces the idea of the existence of a strong interaction between this group and the metal surface. The presence of the out-of-plane deformation vibration of this group as a weak band at $767\,cm^{-1}$ further supports this assumption. The high intensity of the SERS bands at 1405 and $1477\,cm^{-1}$ given by the deformation vibrations of the CH_3 groups found in the vicinity of the carboxylate entity demonstrates the existence of a metal-carboxylate group interaction. By taking into account all these considerations, it becomes obvious that the KBP species are adsorbed on the silver surface through the carboxylate group.

The $1001\,cm^{-1}$ band from the SERS spectrum assigned to the stretching vibrations of beta-lactam and thiazolidine rings ($C_{6,7} + C_3N_4C_7$ stretch) is not shifted in comparison with the corresponding Raman band, while its intensity dramatically decreases. This behavior confirms the existence of a large distance between these rings and the silver surface and, having in mind the geometry of the KBP molecule (Fig. 5.4), it suggests the perpendicular or tilted orientation of these rings with respect to the silver surface (Iliescu et al. 2006).

The very weak intensities of the SERS bands attributed to the symmetric and asymmetric stretching vibrations of the CH groups from phenyl and beta-lactam rings ($2938-3066\,cm^{-1}$) are another consequence of the relatively large distance of these rings relative to the metal surface. The enhancement of the band assigned to the phenyl ring stretching vibration and its appearance at the same wavenumber value ($1600\,cm^{-1}$) in the SERS spectrum compared to the Raman band prove again that this ring is not in the proximity of the metal surface and has approximately

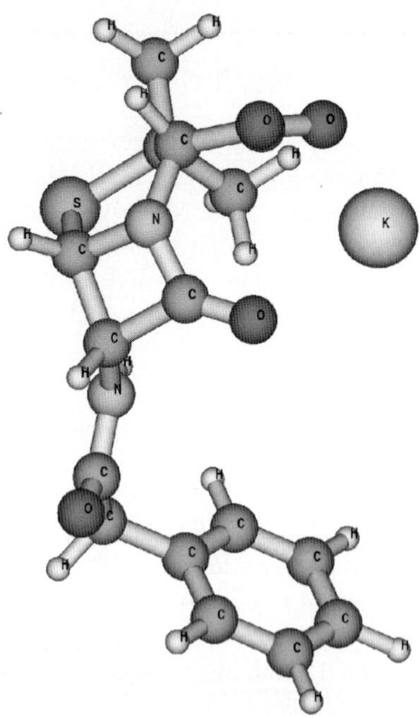

Fig. 5.4 Optimized geometry of the KBP molecule (Raman and SERS investigations of potassium benzylpenicillin, Iliescu T, Baia M, Pavel I, copyright 2006 John Wiley & Sons Limited. Reproduced with permission)

perpendicular orientation with respect to the silver surface. This observation serves as an argument in favor of the electromagnetic enhancement mechanism contribution for these components of the KBP molecule.

In order to understand the adsorption behavior of the KBP species on the metal surface and to get insights about the SERS enhancement mechanisms, electronic absorption spectra of the KBP solution, the pure colloid, and the mixture of the activated silver colloid and the KBP solution have been recorded and are presented in Fig. 5.5.

As one can see, the free KBP species have an absorption maximum in the UV spectral range (Fig. 5.5a) and thus no resonant contribution is expected to the enhancement of the SERS bands. The spectrum of pure silver colloid (Fig. 5.5b) shows a single absorption maximum at 448 nm given by the small particle plasma resonance. The addition of KBP to the activated colloid (Fig. 5.5c) determines a significant decrease of this absorption band and the appearance of a new broad absorption signal at longer wavelength values (around 900 nm). The later absorption peak is known to arise from the aggregation of the colloid particles formed upon an addition of the adsorbed molecule (Blatchford et al. 1982). The major changes evidenced between the Raman and SERS spectra of the KBP molecule, corroborated with the features of the absorption spectrum of the activated silver colloid with added adsorbents, indicated the chemisorption of the KBP species on the colloidal silver particles and the contribution of the charge-transfer effect to

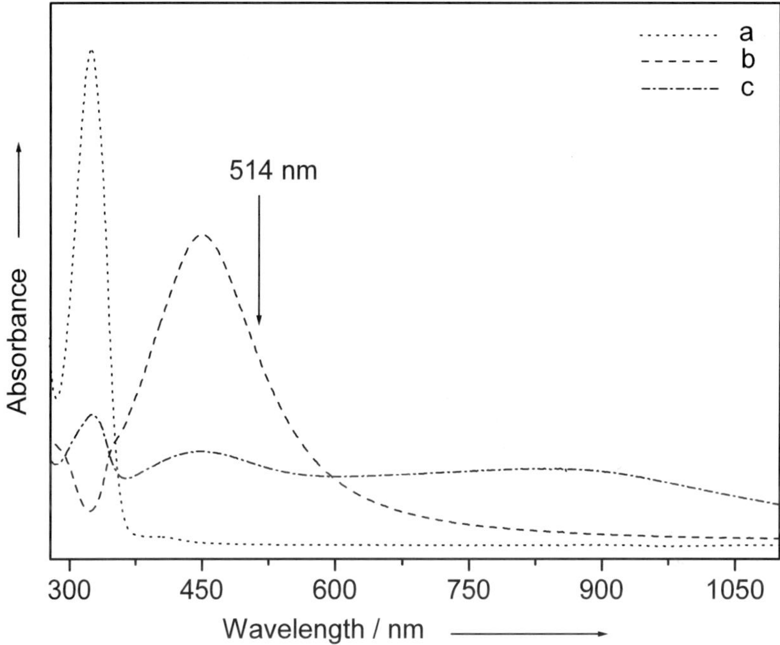

Fig. 5.5 Absorption spectra of KBP 5×10^{-1} M water solution (*a*), pure silver colloid (*b*), and with 5×10^{-1} M KBP solution and 10^{-1} M NaCl (*c*) (Raman and SERS investigations of potassium benzylpenicillin, Iliescu T, Baia M, Pavel I, copyright 2006 John Wiley & Sons Limited. Reproduced with permission)

the overall SERS enhancement. However, for some groups of the KBP molecule, such as phenyl, beta-lactam, and thiazolidine rings, which are situated at a relatively large distance from the metal surface, the electromagnetic enhancement can be considered as the main mechanism of enhancement (Iliescu et al. 2006).

5.1.3 Conclusions

Vibrational investigations have been performed on the KBP molecule by using Raman spectroscopy in conjunction with DFT calculations. Theoretical values of the molecule structural parameters agree well with experimental X-ray diffraction data. The assignment of the vibrational modes was accomplished on the basis of the results derived from theoretical calculations and a fairly good agreement between the theoretical and experimental data was obtained.

The analysis of the SERS spectrum of the KBP recorded on the silver colloid near neutral environment revealed the chemisorption of the KBP molecule via the carboxylate anion. The adsorption of the KBP molecule on the silver surface takes place in such a way that the phenyl, beta-lactam, and thiazolidine rings are located at a relatively large distance to the silver surface and are oriented approximately perpendicularly to it. For these groups, the electromagnetic mechanism seems to be the main enhancement mechanism.

5.2 Trihydrate Amoxicillin

Amoxicillin is an antibiotic belonging to the large class of penicillins. It is a moderate-spectrum β-lactam antibotic used to treat bacterial infections caused by susceptible microorganisms (Budhani and Struthers 1998, Feder Jr. et al. 1999). Amoxicillin is usually the drug of choice within the class because it is better absorbed through oral administration than other beta-lactam antibiotics. It is susceptible to degradation by β-lactamase producing bacteria, and so may be combined with clavulanic acid, a β-lactamase inhibitor, to increase the spectrum of action against Gram-negative organisms, and to overcome bacterial antibiotic resistance mediated through β-lactamase production.

Like other penicillin derivatives, amoxicillin acts by inhibiting the synthesis of bacterial cell walls (Walsh 2000). It inhibits cross-linkage between the linear peptidoglycan polymer chains that make up a major component of the cell wall of Gram-positive bacteria.

When the organism in a serious infection cannot be isolated, a common strategy is to attempt to cover for all possible bacteria. Amoxicillin is frequently used in combination with other antibiotics for this purpose.

In this study, trihydrate amoxicillin (THA) (see Fig. 5.6) was investigated from an experimental (Raman and SERS) and theoretical (DFT) point of view. The results of DFT calculations were compared with the experimental X-ray and Raman data. The spectra obtained from THA in powder form, in a solution and adsorbed on colloidal silver particles have been analyzed in order to elucidate the adsorption behavior of this molecule on the silver surface. This analysis may be regarded as a starting point for further medical studies aimed to identify the most favorable strategies for particular infections.

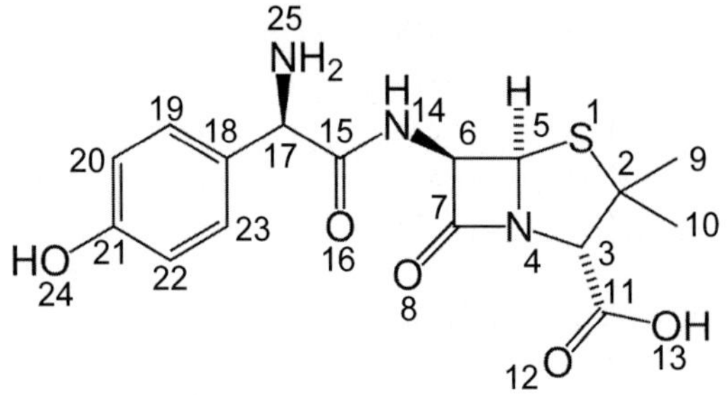

Fig. 5.6 Schematic structure of the trihydrate amoxicillin salt with the labeling of the atoms

5.2.1 *Vibrational Analysis*

The schematic structure of THA with the labeling of the atoms is illustrated in Fig. 5.6.

DFT calculations have been accomplished on THA by using the B3LYP method in conjunction with the split valence shell 6-31G* basis set. The selected calculated structural parameters are given in Table 5.4, together with the available X-ray data (Ghassempour et al. 2007). As one can see from Table 5.4, the calculated bond lengths and bond angles are in agreement with the experimental X-ray data, the observed difference between dihedral angles being most probably due to the intermolecular interactions, which occur in the crystal between amoxicillin and the water molecule.

Table 5.4 Selected calculated bond lengths (Å) and angles (degree) of THA compared with the experimental data

	Calc.[a]	Exp.[b]
Bond lengths (Å)		
S_1-C_2	1.879	1.845
C_2-C_3	1.586	1.556
C_3-N_4	1.452	1.458
N_4-C_5	1.471	1.368
C_5-S_1	1.839	1.779
C_2-C_9	1.539	1.513
C_2-C_{10}	1.533	1.504
C_3-C_{11}	1.527	1.547
C_{11}-O_{12}	1.206	1.221
C_{11}-O_{13}	1.358	1.257
C_5-C_6	1.568	1.519
C_6-C_7	1.552	1.562
C_7-O_8	1.204	1.192
C_7-N_4	1.402	1.495
C_{18}-$C_{19\ average}$	1.396	1.374
C_6-N_{14}	1.427	1.425
N_{14}-C_{15}	1.370	1.344
C_{15}-O_{16}	1.225	1.222
C_{15}-C_{17}	1.538	1.517
C_{17}-C_{18}	1.530	1.511
Angles (degree)		
S_1-C_2-C_3	104.3	104.9
C_3-N_4-C_5	120.8	117.4
N_4-C_5-$S_{1'}$	112.9	103.4
C_5-N_4-C_7	63.1	46

Table 5.4 (Continued)

	Calc.[a]	Exp.[b]
C_2-C_3-C_{11}	111.9	114.2
C_3-C_{11}-O_{12}	126.3	114.8
C_{17}-C_{15}-O_{16}	121.9	121.4
O_{12}-C_{11}-O_{13}	123.4	126
C_5-C_6-C_7	67.7	84.7
C_6-C_7-O_8	136.8	135.5
O_8-C_7-N_4	131.3	130.7
C_{18}-C_{17}-N_{25}	115.8	111.3
C_{18}-C_{19}-$C_{20average}$	120.2	119.9
C_5-C_6-N_{14}	120.8	114.6
C_6-N_{14}-C_{15}	122	123
C_{20}-C_{21}-O_{24}	131.6	116.6
N_{14}-C_{15}-O_{16}	130.3	124.6
C_{17}-C_{18}-C_{19}	120.8	121.5
Dihedral angles (degree)		
S_1-C_5-N_4-C_3	153.4	157.3
S_1-C_5-N_4-C_7	103.6	109.2
S_1-C_5-C_6-N_{14}	154.8	162.5
C_6-C_7-N_{14}-C_{15}	−143.8	−129.2
C_2-C_3-C_{11}-O_{12}	100.4	118
O_8-C_7-N_4-C_3	−34.7	−37.3
O_{16}-C_{15}-N_{14}-C_6	1.3	1
C_{19}-C_{20}-C_{21}-C_{24}	−179.9	−178.9

Abbreviations: [a] Calculated with B3LYP/6-31 + G*, [b] Ref. (Ghassempour et al. 2007)

Figure 5.7 presents the Raman spectrum of polycrystalline THA together with the theoretical spectrum obtained from DFT calculations.

The observed experimental and theoretical Raman bands together with their assignment are summarized in Table 5.5. As can be seen from Table 5.5, the theoretical calculations match well with the experimental data and allow the assignment of the vibrational modes.

In Fig. 5.7, one can see that the most intense bands appears at 2969 (calc. 2945 cm^{-1}) and 853 cm^{-1} (calc. 851 cm^{-1}) and are determined by the stretching vibrations of CH bonds and the in-plane deformation vibrations of the beta-lactam and thiazolidine rings. Other intense Raman bands, which are due to the bending vibrations of the CH and OH groups, can be observed at 1196 (calc. 1185 cm^{-1}) and 1258 cm^{-1} (calc. 1260 cm^{-1}), respectively. The phenyl ring breathing vibration gives rise to the medium-intense band at 986 cm^{-1} (calc. 1000 cm^{-1}), while the in-plane deformation vibration of both the beta-lactam and the thiazolidine ring contribute to the intensity of the band from 853 cm^{-1} (calc. 851 cm^{-1}).

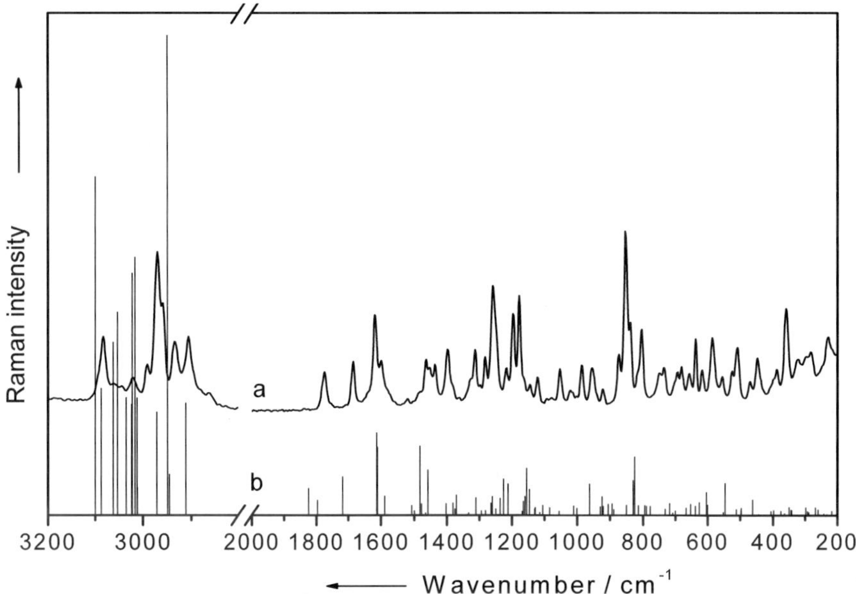

Fig. 5.7 Raman spectra of THA: solid spectrum (*a*) and theoretical spectrum (*b*)

Table 5.5 Raman bands (experimental and calculated) of THA with their assignment

Raman solid sample	Solution	Calc.	Vibrational assignment
228 w	237 sh	234	CH_3 rock + C_7O_8 wag + ring1 in-plane def
282 w	–	288	CH_3 rock + $N_{14}H$ bend + $N_{25}H_2$ rock + ring 2 in-plane def
298 sh	–	298	CH_3 rock
359 s	–	364	CH_3 rock + COOH def
386 w	391 m	376	CH wag (ring 1) + $N_{25}H_2$ rock + $N_{14}H$ wag + COOH def
398 w	–	398	$N_{14}H$ wag + CH def (ring 1) + $O_{13}H$ def + $N_{25}H_2$ twist + ring 2 in-plane def
448 m	–	443	$N_{14}H$ wag + CH_3 rock + ring 2, 3 in-plane def
469 w	–	462	$N_{14}H$ wag
–	485 sh	497	$N_{14}H$ wag + $O_{13}H$ def + ring 3 in-plane def
508 m	–	512	Ring1 out-of-plane + $N_{14}H$ wag + ring 2, 3 in-plane def + CH_3 rock + COOH def
554 w	–	551	CH def (ring 1) + $N_{25}H_2$ wag + $O_{13}H$ def
585 m	–	600	Ring 2 out-of-plane + $O_{13}H$ def
–	603 sh	604	$O_{13}H$ def + CH_3 rock + ring3 in-plane def
616 w	–	626	Ring 1 in-plane def
637 m	–	638	$O_{13}H$ def + CH_3 def
656 w	–	653	$O_{13}H$ def + ring 2, 3 in-plane def

Table 5.5 (Continued)

Raman		Calc.	Vibrational
solid sample	Solution		assignment
693 w	697 w	700	Ring1 out-of-plane + $N_{25}H_2$ twist + CH wag
735 w	–	732	Ring1 in-plane def + $N_{25}H_2$ def + $O_{13}H$ def
803 m	–	795	CH bend (ring 1) + $N_{25}H_2$ wag + $N_{14}C_{15}C_{17}$ bend
838 sh	–	830	CH bend (ring 1) + $N_{25}H_2$ def + ring2 in-plane def
853 vs	–	851	Ring 2, 3 in-plane def + $C_5C_6C_7$ stretch + $O_{13}H$ def
872 w	879 w	889	$N_{25}H_2$ wag + $C_{17}H$ def + CH bend (ring 1)
921 w	–	920	CH bend (ring 1)
953 w	–	962	CH bend (ring 2)
986 mw	–	1000	Ring 1 breathing
1021 w	1018 vs	1011	CH_3 wag + C_6C_7 stretch + C_3H def + ring2 in-plane def
1052 w	–	1055	C_3H def + CH bend (ring 2) + C_7N_4 stretch + $N_{25}H_2$ wag
1081 sh	1078 m	1084	CH def (ring 1) + $N_{25}H_2$ wag
1121 w	1118 w	1128	CH_3 def + C_5H def
1145 w	–	1146	CH def (ring 2)
1177 s	–	1181	$C_{17}H$ def + C_5H def + $N_{14}H$ def + $N_{25}H_2$ twist
1196 s	–	1185	CH_3 bend + $O_{13}H$ def
1216 w	–	1211	CH def (ring 2) + ring 2 in plane def. + $N_{25}H_2$ twist
–	1234 w	1225	$N_{25}H_2$ twist + $N_{14}H$ def + $C_{17}H$ def
1258 s	–	1260	CH bend (ring 3) + $O_{13}H$ bend
1281 w	1280 sh	1281	Ring 2 in plane def + CH bend (ring 2) + $O_{13}H$ def
1312 m	–	1310	CH bend (ring 2) + $N_{14}H$ bend + $N_{25}H_2$ twist + ring 2 in plane def.
1324 sh	–	1332	OH bend + CH rock (ring1)
1378 sh	–	1370	$N_{25}H_2$ def + $C_{17}H$ def
1397 m	–	1402	CH_3 bend
1435 m	–	1436	OH bend + CH rock (ring1) + C_{17},H def
1453 sh	1452 s	1457	CH_3 bend
1462 m	–	1465	
1480 sh	–	1482	
1518 w	–	1507	$N_{14}H$ bend + CH rock (ring1) + $C_{15}N_{14}$ stretch
1599 m	1581 m	1589	CC stretch (ring 1)
1619 s	–	1613	CC stretch + $N_{25}H_2$ wag
1684 m	–	1719	$C_{15}O_{16}$ stretch + $N_{25}H_2$ bend + $N_{14}H$ bend
1775 m	–	1797	$C_{11}O_{12}$ stretch
2969 m	–	2945	CH stretch

Abbreviations: [a]Calculated with B3LYP/6-31 + G*, v = very; w = weak; m = medium; s = strong; sh = shoulder; def = deformation, rock = rocking; bend = bending; wag = wagging; twist = twisting; stretch = stretching; ring 1 = phenyl ring, ring 2 = beta-lactam ring, ring 3 = thiazolidine ring

The band due to the stretching vibrations of the CC groups from the phenyl ring and the deformation vibrations of the amino group is observed at $1619\,cm^{-1}$ (calc. $1613\,cm^{-1}$). The amide I band ($C_{15}O_{16}$ stretching mode) appears in the Raman spectrum at $1684\,cm^{-1}$ (calc. $1719\,cm^{-1}$), while the amide II band is located at $1518\,cm^{-1}$ (calc. $1507\,cm^{-1}$).

The thiazolidine ring deformation vibrations are present in the spectrum at 448 (calc. $443\,cm^{-1}$), 508 (calc. $511\,cm^{-1}$), and $656\,cm^{-1}$ (calc. $653\,cm^{-1}$), while the band due to the deformation vibrations of the beta-lactam ring appears around 398 (calc. $398\,cm^{-1}$), 585 (calc. $600\,cm^{-1}$) and $838\,cm^{-1}$ (calc. $830\,cm^{-1}$).

5.2.2 Adsorption on the Silver Surface

The SERS spectrum of THA at the pH value of 6 together with its solid state Raman spectrum are illustrated in Fig. 5.8.

The pH value of 6 for the colloidal suspension was selected to be similar to that of the human body. Also, is known that in acidic and basic environments, the THA molecule can be broken (Liese et al. 2001). The observed SERS and Raman bands, together with their assignment, are summarized in Table 5.6.

As revealed by Fig. 5.8, the SERS spectrum exhibits a background signal in the spectral range between 1200 and $1600\,cm^{-1}$, which is probably due to the photo or thermal decomposition of the THA molecule that forms carbon layers on the silver nanoparticles surface (Oh et al. 1991).

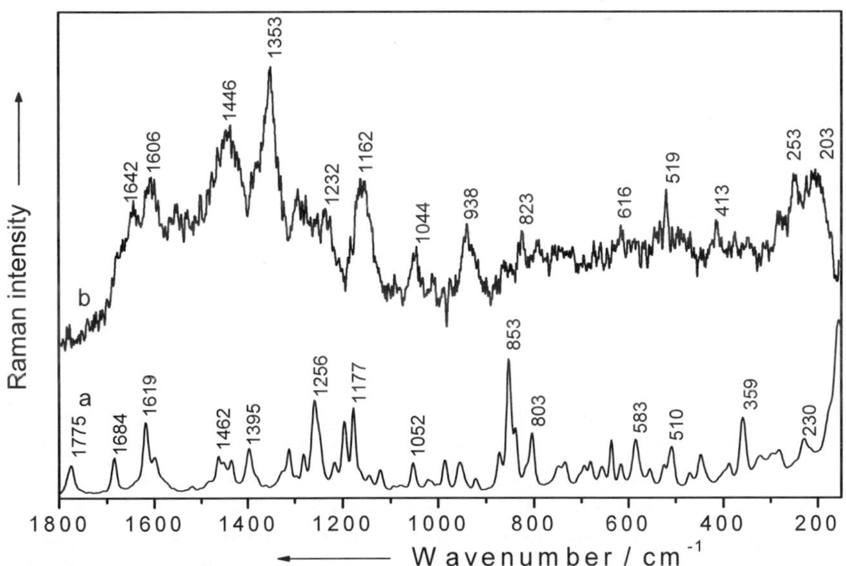

Fig. 5.8 Raman spectrum of polycrystalline THA (*a*) and SERS spectrum of THA at the pH value of 6 (*b*)

Table 5.6 Wavenumbers (cm^{-1}) and assignment of vibrational modes of THA molecule to the SERS bands at the pH value of 6

Raman	SERS pH 6	Vibrational assignment
228 w	253 s	CH$_3$ rock + C$_7$O$_8$ wag + ring1 in-plane def
282 w	284 m	CH$_3$ rock + N$_{14}$H bend + N$_{25}$H$_2$ rock + ring 2 in-plane def
298 sh	308 w	CH$_3$ rock
359 s	374 w	CH$_3$ rock + COOH def
508 m	519 m	Ring1 out-of-plane + N$_{14}$H wag + ring 2, 3 in-plane def + CH$_3$ rock + COOH def
616 w	616 w	Ring 1 in-plane def
803 m	793 w	CH bend (ring 1) + N$_{25}$H$_2$ wag + N$_{14}$C$_{15}$C$_{17}$ bend
838 sh	823 w	CH bend (ring 1) + N$_{25}$H$_2$ def + ring2 in-plane def
872 w	863 w	N$_{25}$H$_2$ wag + C$_{17}$H def + CH bend (ring 1)
953 w	938 m	CH bend (ring 2)
1021 w	1012 w	CH$_3$ wag + C$_6$C$_7$ stretch + C$_3$H def + ring2 in-plane def
1052 w	1044 m	C$_3$H def + CH bend (ring 2) + C$_7$N$_4$ stretch + N$_{25}$H$_2$ wag
1081 sh	1088 w	CH def (ring 1) + N$_{25}$H$_2$ wag
1145 w	1162 s	CH def (ring 2)
1258 s	1232 m	CH bend (ring 3) + O$_{13}$H bend
1281 w	1291 m	Ring 2 in plane def + CH bend (ring 2) + O$_{13}$H def
1378 sh	1353 vs	N$_{25}$H$_2$ def + C$_{17}$H def
1435 m	1446 s	OH bend + CH rock (ring1) + C$_{17}$H def
1599 m	1606 s	CC stretch (ring 1)
1684 m	1642 w	C$_{15}$O$_{16}$ stretch + N$_{25}$H$_2$ bend + N$_{14}$H bend
1775 m	1746 w	C$_{11}$O$_{12}$ stretch

Abbreviations: v = very; w = weak; m = medium; s = strong; sh = shoulder; def = deformation, rock = rocking; bend = bending; wag = wagging; twist = twisting; stretch = stretching; ring 1 = phenyl ring, ring 2 = beta-lactam ring, ring 3 = thiazolidine ring

By inspecting Fig. 5.8, several changes in the relative intensities of some SERS bands in comparison with their analogue Raman bands can be observed. Hereby, the bands around 1350, 1446, and 1160 cm^{-1} become the most intense in the SERS spectrum, while the intensity of the band located at 863 cm^{-1} is very weak in comparison to the corresponding band from the bulk Raman spectrum. Moreover, shifts of some SERS bands compared to their corresponding Raman bands can also be seen. These demonstrate the chemisorption process of the THA molecule on the colloidal nanoparticles surface through some of its constituent groups, while other groups are situated at a relatively large distance from the metal surface, their vibrations being slightly influenced by adsorption. For example, the weak intensity of the SERS bands at 616 and 519 cm^{-1} specific to the rings vibrations proves that these rings are located at a large distance from the metal surface.

The very strong intensity of the band at 1353 cm^{-1} due to the amino group vibration indicates that the THA molecule is bonded to the silver surface by this group.

An evaluation of the adsorbate orientation, and hence the surface binding geometry of THA, can be obtained by using the surface selection rules (Creighton 1983, Moskovits 1985). Having in mind the THA molecule structure, it is obvious that the bonding of this species to the silver surface can be accomplished through different atoms. The absence in the SERS spectrum of the bands due to vibrations in which the sulfur atom from the thiazolidine ring is involved (Table 5.5) proves that this atom is not involved in the THA adsorption on the silver surface. The very strong intensity of the 1353 cm^{-1} band in the SERS spectrum indicates that the $N_{25}H$ group lies in the proximity of the silver surface. On the other hand, the very weak intensity or the absence of the bands due to $N_{14}H$ group vibration is a proof that this group is not involved in the molecule adsorption on the silver surface.

The presence of the carbonyl group ($C_{15}O_{16}$) in the vicinity of the amino $N_{25}H_2$ group (Fig. 5.6) is an indication of the possible involvement of this group in the molecule adsorption. However, because of the very weak intensity of the carbonyl group bands, one can infer that the adsorption of THA molecule takes place most probably by the NH_2 amino group.

The absence or the very weak intensity in the SERS spectrum of the bands due to vibrations, in which the COOH group is involved (374 and 519 cm^{-1}), also demonstrates that this group is situated at a relatively large distance from the silver surface.

By taking into account all these considerations, it becomes obvious that the THA species are adsorbed on the silver surface through the amino group. Figure 5.9 illustrates the optimized geometry of the THA molecule with the most probable adsorption geometry.

The electronic absorption spectra of the pure THA solution, the silver colloid, and the mixture of the activated silver colloid with the THA solution were recorded

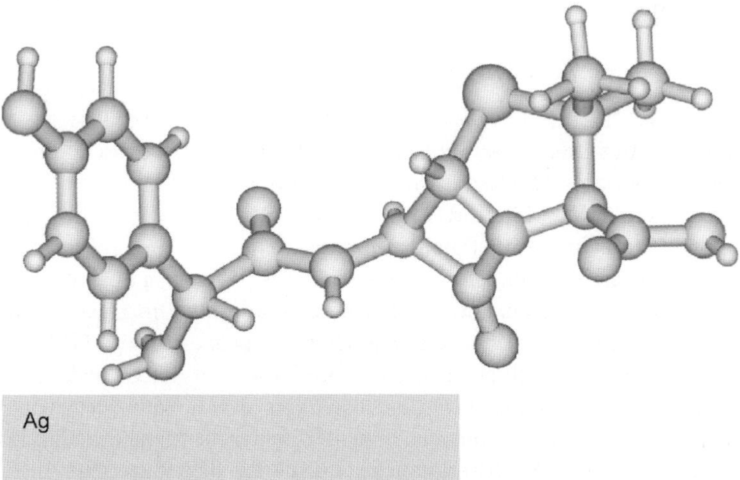

Fig. 5.9 Schematic model for the adsorption geometry of THA on a colloidal silver surface

and analyzed to get further insights about the molecule's adsorption behavior (spectra not shown). It was found that the UV-vis absorption spectrum of the activated silver colloid with THA solution shows, besides the absorption peak at 423 nm given by the small particle plasma resonance, a new broad absorption band around 800 nm, due to the aggregation of the colloidal particles. Thus, the most important changes evidenced between the Raman and SERS spectra of the THA molecule, corroborated with the features of the electronic absorption spectrum, demonstrate the chemisorption of the THA species on the colloidal silver particles and the contribution of the charge-transfer effect to the overall SERS enhancement.

5.2.3 Conclusions

In the study described in this section, Raman spectroscopic investigations, in combination with DFT calculations, have been accomplished on the THA molecule. The theoretical values of the molecule structural parameters reproduce well the experimental X-ray diffraction data. The assignment of the vibrational modes was performed on the basis of the results of theoretical calculations and a fairly good agreement between experimental and theoretical data was achieved.

The analysis of the SERS spectrum of THA recorded on a silver colloidal suspension at the pH value of 6 evidenced the chemisorption process of the THA molecules, which are bonded on the silver surface through the amino group. It was found that the phenyl, beta-lactam, and thiazolidine rings of the THA species are located at a relatively large distance with respect to the silver surface.

5.3 Rivanol

Rivanol (RIV) is an effective antibacterial and has found long-term use orally in enteric diseases such as traveler's diarrhea and shigellosis because of its poor absorption (Wainwright 2001). It was also widely used as a bactericide in the treatment of bovine streptomastitis (Wilson and Giswald 1962).

As one of the highly basic 9-aminoacridine derivatives, RIV is also potentially interesting as an absorbing and fluorescing probe of the nucleic acid structure and of the interaction of aromatic cations with nucleic acids (Wilson and Giswald 1962).

In this work, the FT-Raman spectrum of the solid polycrystalline RIV was recorded and analyzed in order to accomplish a complete vibrational analysis (reprinted from Talanta, 53, Iliescu T, Cinta S, Kiefer W, FT-Raman and SERS spectra of rivanol in silver sol, 121–124, copyright 2000, with permission from Elsevier). Having in mind that the pH value of 5.5 is close to that existing in living tissues, where RIV acts as an antimicrobian drug, the SERS spectrum of RIV on the silver colloid at this value was recorded to elucidate the nature of the adsorption mechanism of this species. In this investigation the silver surface may be considered as an analogue for an artificial biological interface, and after elucidating the adsorption behavior of RIV, the study can be expanded to the adsorption on membranes or other interesting biological surfaces for medical and therapeutic treatments (Dryhurst 1977).

5.3.1 Vibrational Analysis

RIV has three basic nitrogen atoms, and is capable of participating in three protolytic equilibria involving four distinct species: neutral (N), monocation (M), dication (D), and trication (T) (Naik and Schulmann 1975). These species, together with the pK_a values corresponding to their interconversion, are depicted in Fig. 5.10.

In order to have only one preponderant species, the pH value of 5.5 was chosen, when the concentration of monocation specie is theoretically five orders of magnitude greater than that of the dication or the neutral species. The absorption spectrum of RIV solution at the pH value of 5.5 confirms the preponderance of the monocation species (Naik and Schulmann 1975).

Solid RIV exhibits a large fluorescence at the visible excitation light, and therefore infrared excitation was necessary. The FT-Raman spectrum of RIV is displayed in Fig. 5.11.

The lactate distinct bands (Cassanas et al. 1991) are indicated by an asterisk, the rest of them being covered by the cation bands. In the solid state, the only monocation species is present because of the first possible protonation of the

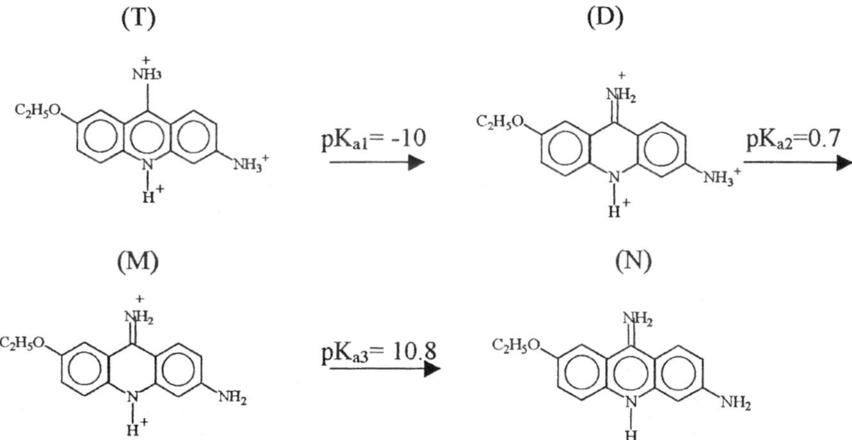

Fig. 5.10 The protolitic equilibrium involving four species of rivanol: T, trication; D, dication; M, monocation; and N, neutral. Reprinted from Talanta, 53, Iliescu T, Cinta S, Kiefer W, FT-Raman and SERS spectra of rivanol in silver sol, 121–124, copyright 2000, with permission from Elsevier

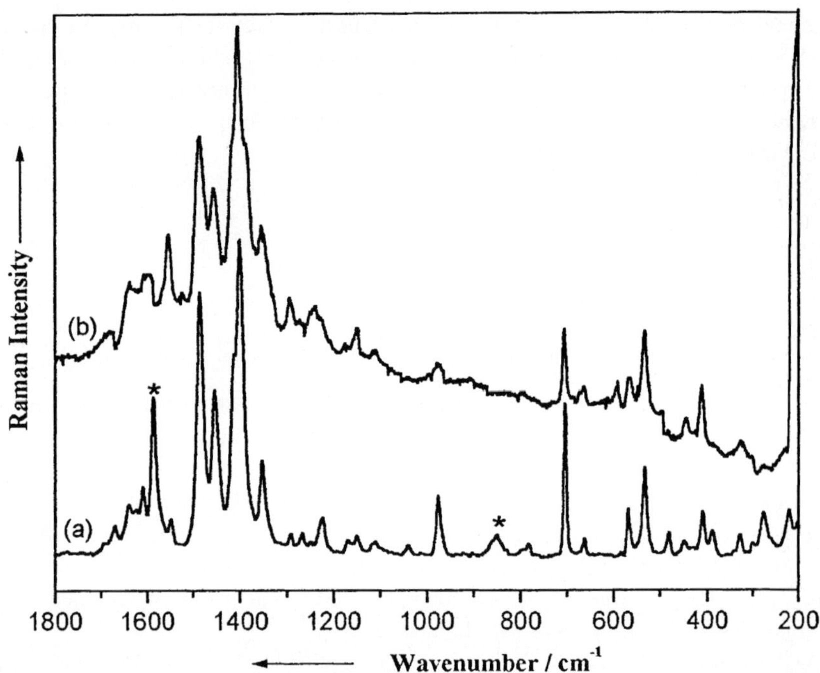

Fig. 5.11 FT-Raman spectrum of RIV in solid state (*a*) and SERS spectrum in silver colloid at a pH value of 5.5 (*b*). Asterisks denote lactate bands. Reprinted from Talanta, 53, Iliescu T, Cinta S, Kiefer W, FT-Raman and SERS spectra of rivanol in silver sol, 121–124, copyright 2000, with permission from Elsevier

nitrogen ring atom, the lactic acid being weak. The observed bands in the FT-Raman spectrum of RIV and their vibrational assignment are summarized in Table 5.7. The proposed assignments were made by a comparison to the acridinium ion (Oh et al. 1991), 9-phenyl- and 9-methyl-acridinium ions (Iliescu etl al. 1994, Iliescu et al. 1995).

Table 5.7 Assignment of the normal vibrational modes of RIV

Raman	SERS pH 5.5	Vibrational assignment
397 vw	–	
417 vw	418 w	Out-of-plane ring def
538 m	538 m	
576 w	580 w	
709 vw	709 m	CH out-of-plane def
787 vw	788 m	NH out-of-plane def
981 w	980 w	
1014 vw	–	
1156 vw	1156 w	CH in-plane def
1176 vw	1177 vw	
1227 w	1229 m	

Table 5.7 (Continued)

Raman	SERS pH 5.5	Vibrational assignment
1297 vw	1300 m	
1359 m	1358 m	Ring stretch
1406 vs	1406 vs	
1418 m	1420 m	
1460 s	1460 s	
1491 vs	1492 s	
1554 vw	1555 m	
1616 w	1617 m	
1643 w	1644 w	NH in-plane def

Abbreviations: w = weak, m = medium, s = strong, v = very, sh = shoulder, stretch = stretching, def = deformation
Reprinted from Talanta, 53, Iliescu T, Cinta S, Kiefer W, FT-Raman and SERS spectra of rivanol in silver sol, 121–124, copyright 2000, with permission from Elsevier

5.3.2 Adsorption on the Silver Surface

The SERS spectrum in the silver colloid at the pH value of 5.5 is presented in Fig. 5.11b. The low concentration ($8.7 \cdot 10^{-6}$ M) of the utilized RIV to obtain SERS is a proof that there is an enhancement of the Raman signal; the normal Raman spectrum recorded with visible excitation light from solutions of such a concentration cannot be obtained. In the SERS spectrum the fluorescence is quenched, probably due to the nonradiative energy transfer to the silver particles (Moskovits 1985).

By comparing the Raman and SERS spectra of RIV, two observations can be derived. Firstly, one can observe that the peak positions, in the limit of experimental errors, are the same. Secondly, one can notice that the lactate peaks at 1592 and $857 \, cm^{-1}$ are absent in the SERS spectrum. This fact confirms the supposition that only the monocation part of the molecular structure of RIV is adsorbed on the silver surface (Iliescu et al. 2000).

It is known that the molecules can adsorb on the metal surface either by physisorption or by chemisorption. The spectrum of physisorbed molecules is practically the same as that of the free molecules, small changes being observed only for the bandwidths (Moskovits 1985). This situation corresponds to a relatively larger distance between the metal surface and absorbed molecules. On the other hand, when the molecules are chemisorbed (Campion and Kambhampati 1998), there is an overlapping of the molecular and metal orbitals, the molecular structure being changed, and, in consequence, the position of the bands and their relative intensities are dramatically changed.

Thus, by taking into account the experimental evidence according to that the SERS and Raman bands of RIV (Fig. 5.11) occurs at the same wavenumber values, one can conclude that there is a physisorption of RIV monocation species on the silver surface. The increased bandwidths observed in the SERS spectrum are probably due to additional vibrational relaxation caused by the interaction between the metal surface and RIV molecule. Because the physisorption of the RIV species was supposed to take place, one can assume that the enhancement of the Raman signal is mainly due to the electromagnetic mechanism (Moskovits 1985, Iliescu et al. 1995). Moreover, it is possible to have some resonance Raman contribution to the total enhancement, as long as the excitation wavelength of 514.5 nm falls on the wing of the RIV absorption band.

5.3.3 Conclusions

The vibrational analysis of rivanol molecules has been performed by using Raman spectroscopy, and it was found that in the solid state sample only the monocation species is present because of the first possible protonation of the nitrogen ring atom. SERS spectra have been also recorded at the pH value of 5.5 and their analysis revealed the physisorption of rivanol monocation species on the silver surface.

References

Blatchford CG, Campbell JR, Creighton JA (1982) Plasma resonance-enhanced Raman scattering by adsorbates on gold colloids: the effects of aggregation. Surf Sci 120:435–455

Budhani RK, Struthers JK (1998) Interaction of Streptococcus pneumoniae and Moraxella catarrhalis: Investigation of the indirect pathogenic role of β-lactamase-producing Moraxellae by use of a continuous-culture biofilm system. Antimicrob Agents Chemother 42:2521–2526

Campion A, Kambhampati P (1998) Surface-enhanced Raman scattering. Chem Soc Rev 27:241–250

Cassanas G, Morssli M, Fabregue E, Bardet L (1991) Vibrational spectra of lactic acid and lactates. J Raman Spectrosc 22:409–413

Coleman K, Athalye M, Clancey A, Davison M, Payne DJ, Perry CR, Chopra I (1994) Bacterial resistance mechanisms as therapeutic targets. J Antimicrobial Chemother 33:1091–1116

Creighton JA (1983) Surface Raman electromagnetic enhancement factors for molecules at the surface of small isolated metal spheres: the determination of adsorbate orientation from SERS relative intensities. Surf Sci 124:209–219

Dexter DD, van der Veen JM (1978) Conformations of penicillin G: crystal structure of procaine penicillin G monohydrate and a refinement of the structure of potassium penicillin G. J Chem Soc Perkin Trans 1:185–190

Dryhurst CG (1977) Electrochemistry of biological molecules. Academic Press, New York

Feder HM Jr, Gerber MA, Randolph MF, Stelmach PS, Kaplan EL (1999) Once-daily therapy for Streptococcal Pharyngitis with amoxicillin. Pediatrics 103:47–51

Gao P, Weaver MJ (1985) Surface-enhanced Raman spectroscopy as a probe of adsorbate-surface bonding: Benzene and monosubstituted benzenes adsorbed at gold electrodes. J Phys Chem 89:5040–5046

Ghassempour A, Rafati H, Adlnasab L, Bashour Y, Ebrahimzadeh H, Erfan M (2007) Investigation of the solid state properties of amoxicillin trihydrate and the effect of powder pH. AAPS PharmSciTech doi: 10.1208/pt0804093

Green M, Wald ER (1996) Emerging resistance to antibiotics: Impact on respiratory infections in the outpatient setting. Annals of Allergy, Asthma, & Immunology 77:167–175

Hu X, Cheng W, Wang T, Wang Y, Wang E, Dong S (2005) J Phys Chem B 109:19385–19388

Iliescu T, Baia M, Pavel I (2006) Raman and SERS investigations of potassium benzylpenicillin. J Raman Spectrosc 37:318–325

Iliescu T, Vlassa M, Caragiu M, Marian I, Astilean S (1995) Raman study of 9-methylacridine adsorbed on silver sol. Vib Spectrosc 8:451–456

Iliescu T, Marian I, Misca R, Smarandache V (1994) Surface-enhanced Raman spectroscopy of 9-phenylacridine on silver sol. Analyst 119:567–570

Iliescu T, Cinta S, Kiefer W (2000) FT-Raman and SERS spectra of rivanol in silver sol. Talanta 53:121–124

Liese A, Seelbach K, Wandrey C (2001) Industrial biotransformation. Wiley, Weinheim

Moskovits M (1985) Surface-enhanced spectroscopy. Rev Mod Phys 57:783–826

Moskovits M, DiLella DP (1980) Surface-enhanced Raman spectroscopy of benzene and benzene-d6 adsorbed on silver. J Chem Phys 73:6068–6075

Naik DV, Schulmann SG (1975) A study of the absorption and fluorescence spectra of rivanol. Anal Chim Acta 80:67–74

Neu H C (1992) The crisis in antibiotic resistance. Science 257:1064–1073

Wainwright M (2001) Acridine- A neglected antimicrobial chromophore. J Antimicrobial Chemotherapy 47:1–13

Walsh CT (2000) Molecular mechanisms that confer antibacterial drug resistance. Nature 406:775–781

Wei A, Kim B, Sadtler B, Tripp S (2001) Chem Phys Chem 12:743–745

Wilson CO, Giswald O (1962) Textbook of organic medicinal and pharmaceutical chemistry. 4th ed. Lippincott, Philadelphia

6 Vitamins

6.1 Vitamin B1

Thiamine (Fig. 6.1), which is also known as vitamin B1, was first isolated in 1926 from rice bran (Schellenberger and Schowen 1988). Thiamine is a water-soluble substance, consisting of a thiazole and pyrimidine-substituted rings joined by a methylene bridge. The human body needs thiamine in order to process carbohydrates, fat, and proteins. Nerve cells require thiamine in order to function normally (Schellenberger and Schowen 1988). Thiamine diphosphate is the active form of thiamine, which serves as a cofactor for several enzymes. The mechanism of action of thiamine enzymes proposed by Breslow (Breslow 1958) 50 years ago is widely accepted. Many details of the catalytic mechanism have been completed by intensive investigations both on model and enzyme systems (Schellenberger and Schowen 1988, Schellenberger 1967, Schowen et al. 1982, Kluger 1987, Jordan et al. 1991, Dyda et al. 1993, Lindquist et al. 1992, Muller and Schulz 1993). Nevertheless, questions concerning this mechanism such as the role of bivalent metal ions Mg^{2+}, Ca^{2+} for the action of thiamine enzymes or the conformation that the whole molecule may take during the enzymatic action are still unanswered today (Malandrinos et al. 2000, Louludi and Hadjiliadis 1994, Friedemann and Neef 1988).

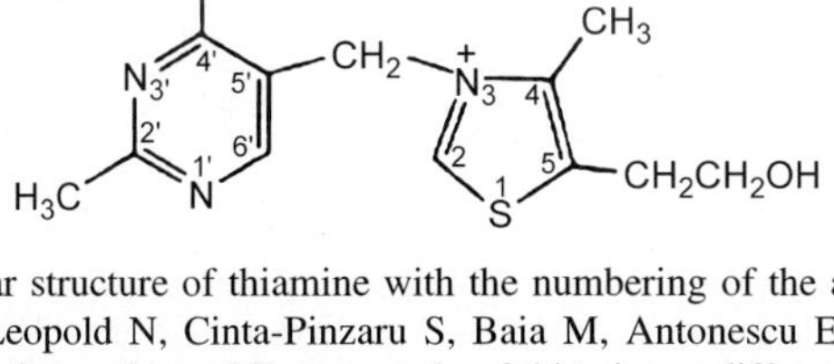

Fig. 6.1 The molecular structure of thiamine with the numbering of the atoms. Reprinted from Vib. Spectrosc., 39, Leopold N, Cinta-Pinzaru S, Baia M, Antonescu E, Cozar O, Kiefer W, Popp J, Raman and surface-enhanced Raman study of thiamine at different pH values, 169–176, copyright 2005, with permission from Elsevier

Several methods such as X-ray diffraction, nuclear magnetic resonance, infrared, and UV-vis spectroscopy were employed in the study of the interaction of bivalent metal ions with thiamine derivatives, and it was found that the metals are bound through the N1' atom and the pyrophosphate group of the phosphate derivatives of thiamine (Malandrinos et al. 2000, Louludi and Hadjiliadis 1994). Ab initio calculations on thiamine systems led to some structural, energetic, and electronic properties of the model system with respect to key steps in the catalytic mechanism (Friedemann and Neef 1988).

A few studies employing vibrational spectroscopic methods for the investigation of thiamine and thiamine derivatives were found in the literature (Butler et al. 1995). Moreover, a SERS study employing a silver electrode for the investigation of thiamine derivatives with pyruvate was also reported by Strekal et al. (Strekal et al. 1992).

Considering the biological importance of thiamine, infrared, Raman, and SERS spectroscopies in combination with DFT calculations were applied to the vibrational characterization of the molecule and are discussed in the next paragraphs. The modifications of the molecular structure of thiamine in different acid and basic aqueous media were evidenced by Raman spectroscopy. The adsorption behavior of the protonated and unprotonated thiamine molecules on the metal surfaces was also monitored by means of SERS spectroscopy (reprinted from Vib. Spectrosc., 39, Leopold N, Cinta-Pinzaru S, Baia M, Antonescu E, Cozar O, Kiefer W, Popp J, Raman and surface-enhanced Raman study of thiamine at different pH values, 169–176, copyright 2005, with permission from Elsevier).

6.1.1 Vibrational Analysis

The possible conformations that thiamine may take are determined by the relative orientations of the thiazole and pyrimidine rings. These orientations are best expressed in terms of the torsion angles about the bonds from the methylene bridge

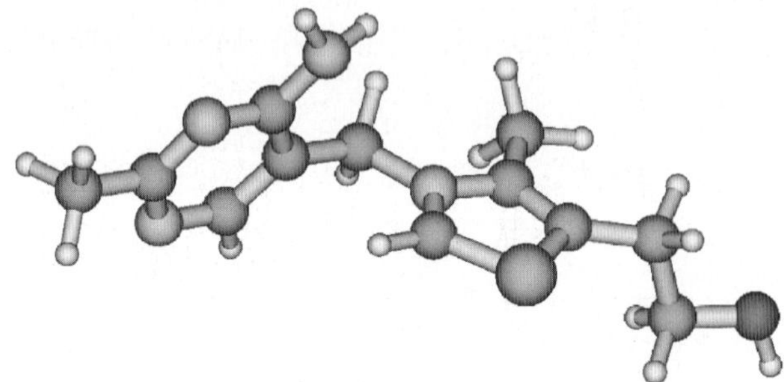

Fig. 6.2 The optimized geometry of the thiamine molecule calculated at the BPW91/6-311 + G* level of theory. Reprinted from Vib. Spectrosc., 39, Leopold N, Cinta-Pinzaru S, Baia M, Antonescu E, Cozar O, Kiefer W, Popp J, Raman and surface-enhanced Raman study of thiamine at different pH values, 169–176, copyright 2005, with permission from Elsevier

carbon to the thiazole and pyrimidine rings, $\Phi_P = N–Cb–C5–C4'$ and $\Phi_T = C5'–Cb–N3–C2$, which have positive values in a clockwise direction (Malandrinos et al. 2000, Shin et al. 1993). Thus, in the F conformation $\Phi_T = \pm 0°$ and $\Phi_P = \pm 90°$, in the S conformation $\Phi_T = \pm 100°$ and $\Phi_P = \pm 150°$ and in the V conformation $\Phi_T = \pm 90°$ and $\Phi_P = \pm 90°$. In numerous crystal structures (Shin et al. 1993, Pletcher et al. 1977), free thiamine assumes mostly the F conformation and with a minor exception, the S conformation, despite the apparent rotational degree of freedom around the bridging methylene joining the two aromatic rings. The average Φ_T and Φ_P angles for the F form were found to be $-4 \pm 7°$ and $83 \pm 6°$ (Shin et al. 1993).

Taking these into consideration, the theoretical calculations on thiamine in the F conformation were performed. The optimized geometry of the thiamine molecule,

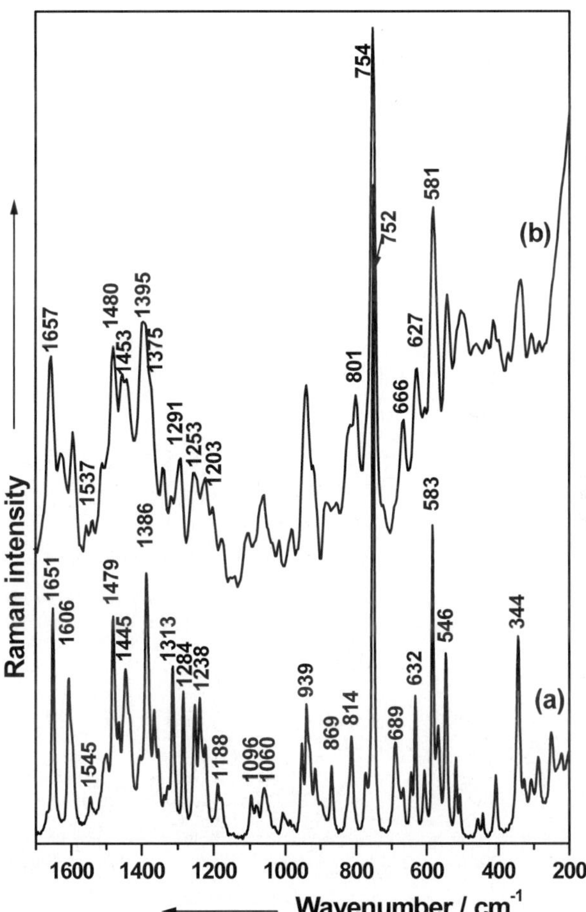

Fig. 6.3 FT-Raman spectra of solid thiamine hydrochloride (*a*) and of its aqueous saturated solution (*b*). Reprinted from Vib. Spectrosc., 39, Leopold N, Cinta-Pinzaru S, Baia M, Antonescu E, Cozar O, Kiefer W, Popp J, Raman and surface-enhanced Raman study of thiamine at different pH values, 169–176, copyright 2005, with permission from Elsevier

calculated at the BPW91/6-311+G* theoretical level, is illustrated in Fig. 6.2. The torsion angles obtained after geometry optimization and wavenumber calculations at the BPW91/6–31+G* and BPW91/6-311+G* levels of theory were found to be $\Phi_T = 4.023°$, $\Phi_P = 78.705°$, and $\Phi_T = 4.798°$, $\Phi_P = 78.240°$, respectively, (Leopold et al. 2005) and are in good agreement with the experimental results (Shin et al. 1993).

Depending on the pH of the aqueous solution, the thiamine molecules exist in protonated or unprotonated (neutral) forms, and the concentration of each molecular species is given by the Henderson–Hasselbach equation (Atkins 1987).

Theoretical calculations (Friedemann and Neef 1988) show that in the gas phase, thiamine is more stable than its imino tautomers and the protonation of N1' atom is energetically preferred with respect to the N3' atom and the N atom of the amine group. Therefore, in the pH range from 1 to 9, only the protonation of the N1' atom is expected (Malandrinos et al. 2000, Louludi and Hadjiliadis 1994, Friedemann and Neef 1988, Strekal et al. 1992).

Figure 6.3 shows the FT-Raman spectra of solid thiamine hydrochloride and its aqueous saturated solution. Thiamine is available as hydrochloride; therefore, the aqueous saturated solution has a pH value around 2. At this pH, thiamine molecules exist in the N1' protonated molecular form (Strekal et al. 1992). The characteristic spectral feature of this molecular form is the intense peak at $1657\,cm^{-1}$, due to the supposition of the N4'H$_2$ bending vibration, as revealed by the theoretical

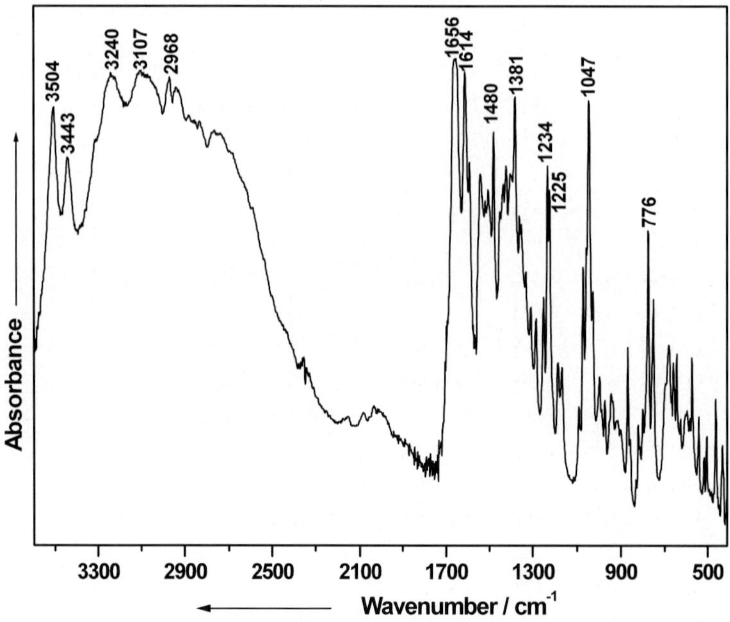

Fig. 6.4 FT-infrared spectrum of solid thiamine hydrochloride. Reprinted from Vib. Spectrosc., 39, Leopold N, Cinta-Pinzaru S, Baia M, Antonescu E, Cozar O, Kiefer W, Popp J, Raman and surface-enhanced Raman study of thiamine at different pH values, 169–176, copyright 2005, with permission from Elsevier

calculations (Table 6.1) and the N1'H bending vibration of the protonated nitrogen, respectively (Socrates 2001). This strong peak is present in the FT-Raman spectrum of solid thiamine hydrochloride at $1651\,cm^{-1}$ (Fig. 6.3a).

The infrared spectrum of thiamine hydrochloride (Fig. 6.4) shows strong bands at 1656, 1614, 1480, 1381, 1234, 1225, 1047, and $776\,cm^{-1}$ in agreement with the previously published data (Schrader 1995).

The main bands of the Raman and infrared spectra of solid thiamine and their assignment carried out with the help of theoretical calculations performed at the BPW91/6-31 + G* and BPW91/6-311 + G* levels of theory are given in Table 6.1.

Table 6.1 Raman and infrared bands (experimental and calculated) of thiamine and their vibrational assignment

Infrared	Raman	Calc.[a]	Calc.[b]	Vibrational assignment
–	221 w	226	225	C2S1C5 out of plane def
–	250 m	263	243	OH def
–	287 m	283	283	C5'CbN3 out of plane def
–	307 w	307	306	CCO def + CC (R6-CH$_3$, R6-CbH$_2$) out of plane def
–	327 sh	327	327	R6 out of plane def
–	344 s	332	332	R5 out of plane def
411 w	408 m	426	424	R6 + R5 out of plane def
430 m	444 w	433	431	
461 m	458 w	476	474	CC rock + CN (R6-CH$_3$, R6-NH$_2$, R6-CbH$_2$) stretch
503 m	507 m	506	503	S1C2N3 + S1C5C4 out of plane def
516 w	518 m	527	527	R6 out of plane def
540 m	546 m	553	552	NH (NH$_2$) out of plane def
573 m	566 m	561	558	
582 sh	583 m	576	576	
595 w	605 m	605	604	C2N3C4 out of plane def + C5'Cb stretch
626 w	632 m	622	621	R6 + R5 out of plane def
644 m	644 m	647	648	R6 out of plane def
659 m	667 m	657	658	R5 out of plane def + CS stretch
681 m	689 m	699	699	C5-CH$_2$-CH$_2$ + C2H out of plane def
752 m	752 vs	755	752	R6 breathing
776 s	773 sh	764	762	CH (CH$_2$-CH$_2$) + C2H out of plane def
800 m	814 m	801	800	R6 + R5 out of plane def
820 m	818 sh	827	826	
868 m	869 m	873	872	CS stretch
899 sh	900 sh	907	902	CH (CH$_2$-CH$_2$) def
918 w	915 m	915	912	C6'H + CbH$_2$ out of plane def
934 sh	939 m	940	932	C6'H + CbH$_2$ out of plane def + R6 stretch
943 m	952 m	955	951	R5 breathing

Table 6.1 (Continued)

Infrared	Raman	Calc.[a]	Calc.[b]	Vibrational assignment
1047 s	1043 sh	1056	1047	CO stretch
1058 sh	1058 m	–	–	CO stretch + h R6stretch
1073 m	1083 m	1091	1089	C5C stretch
1091 m	1096 m	1094	1090	NH (NH2) + CH (CH$_2$-CH$_2$, CH$_3$ R6) def
1168 m	1177 m	1146	1141	C2H bend + C2N3Cb stretch
1185 m	1188 m	1194	1193	C5'Cb stretch
1225 s	1222 m	1219	1218	C6'H + C2H bend + N1'C2'N3' stretch
1234 s	1238 m	1277	1271	
1251 m	1251 m	1278	1278	CH (CH$_2$-CH$_2$ + CbH$_2$) def + C5C4C6 stretch
1284 m	1284 m	1286	1282	CH (CH$_2$-CH$_2$ + CbH$_2$) def + C5C4C6 + CN stretch
1308 m	1313 s	1304	1300	CH (CH$_2$-CH$_2$ + CbH$_2$) def + C5C4C6 stretch
1352 m	1353 sh	1355	1350	CH (CH$_3$ R6) def
1361 m	1365 m	1380	1370	CH (CH$_3$ R5) def
1381 s	1386 s	1405	1398	R6 stretch
1405 m	1403 sh	1406	1406	CH (CH$_2$-CH$_2$) + OH def
1423 m	1431 sh	1435	1427	C2'N1'C6' + C2'N3'C4' stretch
1438 m	1445 m	1449	1441	CH (CH$_3$, CbH$_2$) def
1441 sh	1462 m	1464	1457	
1480 s	1479 s	1485	1479	C2N3C4 stretch
1506 m	1498 m	1499	1491	CH (CH$_2$-CH$_2$) def
1542 m	1545 w	1542	1538	R6 stretch
1592 sh	1597 sh	1575	1583	
1614 s	1606 m	1583	1596	
1656 s	1651 m	1631	1637	NH (NH$_2$) + N1'H protonated bend
2738 m	2733 w	2794	2762	CH (CH$_2$-CH$_2$) stretch
2853 sh	2867 sh	2997	2984	CH (CH$_3$ R5) stretch
2879 sh	2889 sh	3002	2988	CH (CH$_3$ R6) stretch
2949 s	2928 vs	3049	3003	CH (CH$_2$-CH$_2$) stretch
2969 s	2966 s	3074	3037	CH (CbH$_2$) stretch
3101 s	3080 sh	3119	3103	CH (CH$_3$ R5) stretch
3238 s	3209 w	3221	3251	CH (CH$_3$ R6) stretch
3442 s	–	3496	3490	NH (NH$_2$) stretch
3504 s	–	3554	3540	OH stretch

Abbreviations: [a] Calculated with BPW91/6–31 + G*, [b] Calculated with BPW91/6–311 + G*, w = weak, m = medium, s = strong, sh = shoulder, stretch = stretching, def = deformation, bend = bending, R5 = thiazole ring, R6 = pyrimidine ring

Reprinted from Vib. Spectrosc., 39, Leopold N, Cinta-Pinzaru S, Baia M, Antonescu E, Cozar O, Kiefer W, Popp J, Raman and surface-enhanced Raman study of thiamine at different pH values, 169–176, copyright 2005, with permission from Elsevier

The calculated wavenumbers are obtained using the harmonic approximation, whereas the experimental wavenumbers are anharmonic by nature. Nevertheless, the quality of the quantum chemical results at the presented theoretical levels is sufficient to be useful for the assignment of the experimental data. The theoretical results obtained with the $6\text{-}311+G^*$ basis set are in the best agreement with the experimental values.

The physical processes and chemical reactions involved in biological processes are often very sensitive to the concentration of the hydrogen ions of the medium. Having in mind these considerations, the Raman spectra of the thiamine aqueous solution at different pH values were recorded and are shown in Fig. 6.5.

The Raman spectra corresponding to different pH values from the 1–7 pH range reveal the presence of two molecular species: the protonated and the unprotonated ones. According to the changes evidenced in the spectra, mainly for the peaks at 1657 and 1603 cm^{-1}, the crossover between the two molecular forms takes place at the pH value of 5; consequently, the pK$_a$ value of the molecule for N1' protonation is near this pH value. This fact is in good agreement with the literature (Malandrinos et al. 2000, Louludi and Hadjiliadis 1994, Strekal et al. 1992).

In the pH range from 1 to 4, the spectra indicate the presence of the protonated thiamine molecular species in the solution. The characteristic Raman bands of the protonated molecular form are those from 1657 and 1550 cm^{-1}. At pH values of 6 and 7, when only unprotonated thiamine molecules contribute to the Raman spectra, the band at 1657 cm^{-1} is present as a weak shoulder, whereas the band at 1550 cm^{-1} disappears.

The Raman spectrum at pH $=5$ reveals the coexistence of both protonated and unprotonated thiamine molecular forms. The peak present at 1597 cm^{-1} in the Raman spectrum of the protonated thiamine molecules at the pH value of 4, appears at 1599 cm^{-1} in the Raman spectrum at pH $=5$, and it is further shifted and becomes very strong even at higher pH values (1603 cm^{-1}). The shift and also the increased intensity of the peak at 1599 cm^{-1}, assigned to the pyrimidine ring stretching vibration, evidence the presence of unprotonated molecular species at a pH value of 5. In a similar way, the increase in intensity of the shoulder at 1377 cm^{-1}, which becomes a distinct band at 1375 cm^{-1} at a higher pH value, reveals also the presence of the unprotonated molecules at pH values equal with and higher than 5.

Beginning with pH $=5$, the number of protonated molecules exponentially decreases, being in agreement with the Henderson–Hasselbach equation.

The spectra recorded at pH values of 6 and 7 are preponderantly due to the unprotonated molecular form of thiamine. The characteristic spectral features for this molecular species in aqueous solution are two peaks at 1603 and 1375 cm^{-1}. Moreover, in the spectra of the neutral thiamine molecules, the bands located at 417, 548, 592, 639, 678, and 760 cm^{-1} are shifted to higher wavenumber values in comparison with those corresponding to the protonated molecular species at pH $=4$.

At basic pH values over 8, the drastic modifications observed in the Raman spectrum clearly indicate the denaturation of the thiamine structure. According to the literature (Louludi and Hadjiliadis 1994, Clarcke and Gurin 1985) in an alkaline medium thiamine becomes a disulphidic form (oxidated), which, in passing to a thiol form (reduced) is forming a redox system.

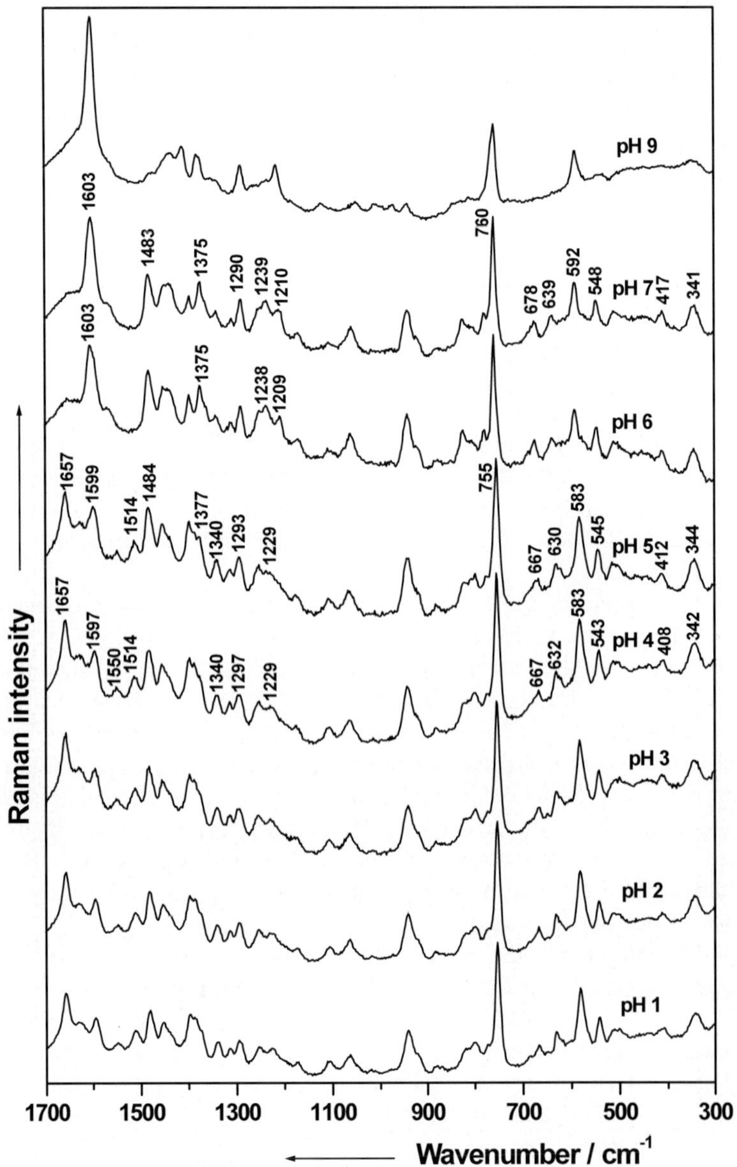

Fig. 6.5 Raman spectra of the thiamine aqueous solution (10^{-1} M) at different pH values as indicated. Reprinted from Vib. Spectrosc., 39, Leopold N, Cinta-Pinzaru S, Baia M, Antonescu E, Cozar O, Kiefer W, Popp J, Raman and surface-enhanced Raman study of thiamine at different pH values, 169–176, copyright 2005, with permission from Elsevier

6.1.2 Adsorption on the Gold Surface

Complementary to Raman spectroscopy, the SERS technique is able to monitor the molecules adsorbed on the metal surface without giving information about the bulk solution. SERS spectra of thiamine on a gold colloidal suspension in the pH range from 1 to 8 are illustrated in Fig. 6.6. The SERS spectra recorded at more alkaline pH values (not showed) reveal the denaturation of the molecule, as it was found from the Raman spectra (Fig. 6.5). The assignment of the SERS bands to their corresponding Raman bands is summarized in Table 6.2.

At pH values of 1 and 2, as it was expected, the spectra reveal the presence of the protonated species adsorbed on the gold surface. As the peaks at 1657 and

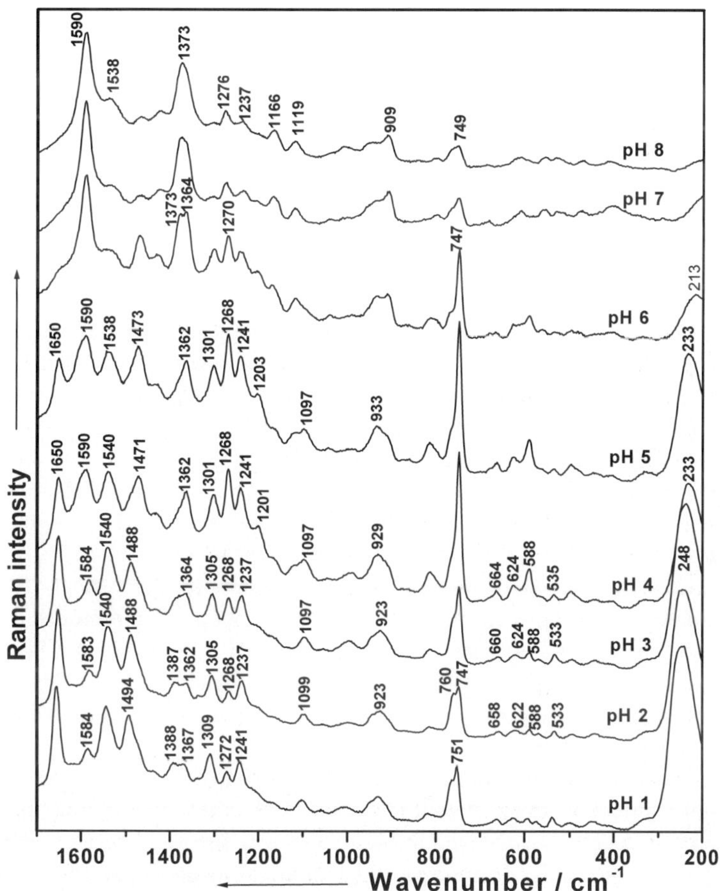

Fig. 6.6 SERS spectra of thiamine in a gold colloidal suspension (10^{-5} M) at different pH values as indicated. Reprinted from Vib. Spectrosc., 39, Leopold N, Cinta-Pinzaru S, Baia M, Antonescu E, Cozar O, Kiefer W, Popp J, Raman and surface-enhanced Raman study of thiamine at different pH values, 169–176, copyright 2005, with permission from Elsevier

Table 6.2 Selected wavenumbers (in cm^{-1}) and the assignment of the vibrational modes of the thiamine molecule to the SERS bands at pH values of 1, 3, and 8

| Raman | SERS | | | Vibrational |
	pH 1	pH 3	pH 8	assignment
–	248 vs	248 s	243	AuCl stretch
250 m	263	–	–	OH def
752 vs	751 s	747 s	749 m	R6 breathing
939 m	923 m	923 m	932 m	C6'H + CbH$_2$ out of plane def
1096 m	1099 m	1097 m	1119 m	NH (NH2) + CH (CH$_2$-CH$_2$, CH$_3$ R6) def
1238 m	1241 m	1237 m	1237 m	C6'H + C2H bend + N1'C2'N3' stretch
1284 m	1272 m	1268 s	1276 m	CH (CH$_2$-CH$_2$ + CbH$_2$) def + C5C4C6 stretch
1313 s	1309 s	1305 s	1301 m	
1386 s	1388 m	1379 m	1373 s	R6 stretch
1498 m	1494	1488	1471 w	CH (CH$_2$-CH$_2$) def
1545 w	1540 s	1540 s	1538 m	R6 stretch
1606 m	1584 m	1584 m	1590 m	R6 stretch
1651 m	1650 s	1650 s	–	NH (NH$_2$) + bend N1'H protonated bend

Abbreviations: w = weak, m = medium, s = strong, sh = shoulder, stretch = stretching, def = deformation, bend = bending, R5 = thiazole ring, R6 = pyrimidine ring
Reprinted from Vib. Spectrosc., 39, Leopold N, Cinta-Pinzaru S, Baia M, Antonescu E, Cozar O, Kiefer W, Popp J, Raman and surface-enhanced Raman study of thiamine at different pH values, 169–176, copyright 2005, with permission from Elsevier

1550 cm^{-1} indicated the presence of the protonated molecules in the Raman spectra (Fig. 6.5) in the same way, the corresponding SERS bands at 1650 and 1540 cm^{-1} evidence the presence of the protonated species adsorbed on the gold surface.

The spectrum recorded at pH = 3 is related to the previous two spectra, but a slight increase in intensity of the SERS band at 1584 and of the shoulder at 1379 cm^{-1}, which is maintained at higher pH values, is observed. These two bands are present at 1590 and 1373 cm^{-1} in the SERS spectrum at a pH value of 8, as the main bands having the corresponding Raman peaks at 1603 and 1375 cm^{-1} in the spectrum at pH 7 (Fig. 6.5), due to the unprotonated molecular species. Therefore, one supposes that at the pH value of 3, a few numbers of unprotonated species coexist with the protonated ones adsorbed on the gold surface. Several SERS studies (Giese and McNaughton 2002a, Giese and McNaughton 2002b) reported the adsorption of the unprotonated molecular form at pH values lower with two or more units than the pK$_a$ of the analyte.

The coexistence of the protonated and unprotonated adsorbed thiamine species on the gold surface is more evident at pH = 4. This is revealed by the presence in the spectra of the bands attributed to both molecular species. The contributions of the protonated molecular form are the bands at 1650 and 1540 cm^{-1}, while the contributions of the unprotonated molecular species are mainly the band at 1590 and the shoulder at 1379 cm^{-1}. On passing from pH 4 to pH 5, the relative intensities of the band at 1590 and the shoulder at 1375 cm^{-1}, due to the neutral thiamine molecules, slightly increase, while those with peaks at 1650 and 1540 cm^{-1}, given by the protonated molecules, decrease. Consequently, an increase of the number of neutral thiamine molecules adsorbed to the gold surface is observed, whereas the number of protonated species decreases. The SERS spectra at pH values of 6, 7, and 8 are preponderantly due to the unprotonated thiamine molecules adsorbed to the gold surface. The characteristic spectral features of these adsorbed molecules are the strong bands at 1590 and 1373 cm^{-1}.

On passing from the Raman spectra (Fig. 6.5) to the SERS spectra (Fig. 6.6), one can see that the wavenumber values of the vibrational modes of the pyrimidine and thiazole rings are gently shifted and the band intensities are enhanced in the studied pH range. For the protonated molecular species, the bands due to the pyrimidine ring vibrations situated in the Raman spectra at 1657, 1550, and 755 cm^{-1} are shifted in the SERS spectra at 1650, 1540, and 747 cm^{-1}, respectively, whereas the Raman bands due to the unprotonated pyrimidine ring vibrations that appear at 1603, 1375, and 760 cm^{-1} are shifted in the SERS spectra at 1590, 1373, and 749 cm^{-1}, respectively. In the same way, when the thiamine molecules are protonated, the Raman bands given by the thiazole ring vibrations from 1251, 1297, and 1484 cm^{-1} are shifted in the SERS spectra at 1268, 1305, and 1488 cm^{-1}, respectively, whereas for the neutral thiamine molecules, the Raman bands of the thiazole ring vibrations that appear at 1255, 1290, and 1483 cm^{-1} are present in the SERS spectra at 1276, 1301 and 1471 cm^{-1}, respectively (Figs. 6.5 and 6.6).

According to the surface-selection rules (Creighton 1983, Moskovits 1984), the normal modes, with a change in the polarizability component perpendicular to the surface, are enhanced. On the other hand, one can assume that the interaction of thiamine with the gold surface can be established through the lone pair electrons of the nitrogen atom(s) from the pyrimidine moiety of thiamine and/or of the S atom from the thiazole, or through the π electrons of the rings. Since the thiazole ring modes are less representative in the SERS spectra, one can suppose that this ring is not directly involved into adsorption even for protonated or neutral forms. The Raman band at 1550 cm^{-1}, attributed to a pyrimidine ring stretching vibration, appears in the SERS spectra at 1540 cm^{-1} and is by far the most enhanced Raman band, in the case of the protonated molecular form. For the neutral molecular form the SERS spectra are dominated by the band at 1603 cm^{-1} assigned also to the pyrimidine ring vibration. Therefore, from the examination of the relative intensities between Raman and SERS spectra, one can suppose that the interaction of thiamine with the gold surface is established through the lone pair electrons of the nitrogen atoms from the pyrimidine ring, for both molecular species (Leopold et al. 2005).

Following the variations of the ring breathing mode of the pyrimidine part of thiamine at around 747 cm^{-1} in the SERS spectra (Fig. 6.6), one can see that this vibration is less enhanced in the spectra recorded at pH values from 1 to 3. Therefore, one assumes that the protonated molecules are adsorbed in a predominantly tilted orientation of the pyrimidine ring to the gold surface only through the N3' atom, the lone pair electrons of the N1' atom being involved in the protonation.

At pH values of 4 and 5, when both molecular species coexist on the gold surface, the ring breathing mode becomes more enhanced; consequently, the thiamine molecules adsorb in a less tilted, closer to perpendicular orientation of the pyrimidine ring with respect to the metal surface. An absolute perpendicular orientation of the pyrimidine ring is hindered by the methyl group, attached to the pyrimidine ring between the N1' and N3' atoms. Consequently, at these pH values, the adsorption of the protonated molecules takes place most probably through the N3' atom, whereas for the unprotonated thiamine molecules an additional adsorption through the N1' atom is also plausible.

At pH values over 6, when the unprotonated thiamine molecules are dominant on the gold surface, the adsorption takes place through the N3' and N1' atoms of the pyrimidine moiety of the thiamine. As can be seen from Fig. 6.6, at pH values of 7 and 8, the pyrimidine ring breathing mode is less enhanced; therefore, a flat orientation of the pyrimidine ring with respect to the surface was supposed (Leopold et al. 2005).

The involvement of the nitrogen atoms in the adsorption of thiamine is further confirmed by the examination of the low wavenumber region of the SERS spectra. Unfortunately, in the SERS spectra at an acidic pH, the AuN stretching vibration is overlapped by the strong AuCl stretching band at around $248 \, cm^{-1}$. However, the band shape indicates at least two contributions. With the pH increasing, the concentration of the Cl^{-} ions is decreased, and consequently, the band profile is dramatically changed, the AuN stretching band at $213 \, cm^{-1}$ becoming observable.

6.1.3 Conclusions

FT-Raman and infrared spectra of thiamine hydrochloride were recorded and the assignment of the vibrational wavenumbers was accomplished with the help of DFT calculations.

The FT-Raman spectra of a solid and aqueous saturated solution of thiamine hydrochloride reveal the presence of the protonated molecular form. The pH-dependence Raman spectra of the thiamine aqueous solution revealed the presence of two molecular species, the protonated and neutral form, respectively, the pK_a value for the protonation of N1' atom being found to be slightly over 5. In a strong alkaline environment (pH > 8) the denaturation of the molecule was observed.

The pH-dependence SERS study revealed the presence of two different adsorbed molecular species and their coexistence. A higher adsorption affinity to the gold surface of the neutral molecular form was concluded, as the adsorption of neutral thiamine species was observed at pH values lower with two units than the pK_a value of the molecule. The chemisorption of the thiamine molecules takes place through the nitrogen atoms of the pyrimidine ring, the orientation of the molecule with respect to the gold surface depending on the protonation degree.

6.2 Vitamin PP

Vitamin PP, also known as nicotinamide, is a water-soluble component of the vitamin B complex group. The PP vitamin is a derivative of vitamin B3 and is important in promoting a healthy nervous system, healthy skin, and proper gastrointestinal functioning.

In cells, the PP vitamin is incorporated into nicotinamide adenine dinucleotide (NAD) and nicotinamide adenine dinucleotide phosphate (NADP), which function as coenzymes in a wide variety of enzymatic oxidation-reduction reactions. The PP vitamin is involved in a wide range of biological processes, including the production of energy, the synthesis of fatty acids, cholesterol, and steroids, signal transduction and the maintenance of the integrity of the genome.

Nicotinamide may have anti-diabetogenic activity in some. It has been used in diabetes treatment and prevention. It may also have antioxidant, anti-inflammatory, and anticarcinogenic activities. The PP vitamin can be used for the treatment of arthritis by aiding the body in its production of cartilage and has putatitive activity against granuloma annulare. Nicotinamide is a potent anti-inflammatory agent used in various dermatological disorders.

In the following paragraphs, the pH influence on the adsorption behavior of the PP vitamin on colloidal silver nanoparticles as evidenced from the SERS spectra is presented. The nature and the orientation of species adsorbed onto metal surface are discussed (reprinted from J. Molec. Struct., 410–411, Iliescu T, Cinta S, Astilean S, Bratu I, pH influence on the Raman spectra of PP vitamin in silver sol, 193–196, copyright 1997, with permission from Elsevier).

6.2.1 Adsorption on the Silver Surface

The change in the structure of the PP vitamin resulting from the variation of the pH is shown in Fig. 6.7.

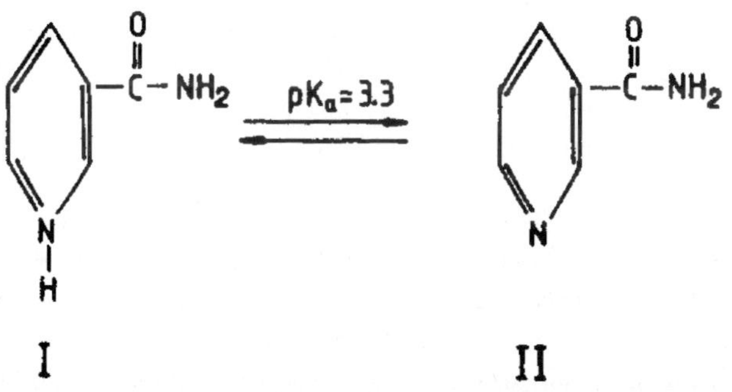

Fig. 6.7 Structural change of the PP vitamin with the variation of the pH. Reprinted from J. Molec. Struct., 410–411, Iliescu T, Cinta S, Astilean S, Bratu I, pH influence on the Raman spectra of PP vitamin in silver sol, 193–196, copyright 1997, with permission from Elsevier

The pK$_a$ value of 3.3 (Grecu and Curea 1980) indicates that at pH values lower than 3.3 the protonated forms (I) are preponderantly present in the solution, while at pH values higher than 3.3 the neutral form (II) of PP vitamin prevails.

The bands appearing in the normal Raman spectrum of solid PP vitamin and in the SERS spectra at pH values of 2.5 and 5.5 are summarized in Table 6.3. The assignment of the vibrational modes was accomplished by comparison with the modes of nicotinic acid (Park et al. 1994, Chang and Furtak 1982, Avram and Mateescu 1966).

Table 6.3 Raman and SERS wavenumbers (in cm^{-1}) of the PP vitamin at different pH values and their assignment

Raman	SERS pH 2.5	pH 5.5	Vibrational assignment
529 w	526 w	527 m	In-plane ring def
772 sh			In-plane ring def
786 m	790 vw	794 m	O=C-NH$_2$
834 w		824 vw	In-plane ring def
	912 vw		Out-of-plane CH def
	968 w		
1037 vs	1024 m	1027 vs	Ring breathing
		1042 sh	
	1070 w		CH bend
1119 w		1124 w	CH bend
1158 w	1160 sh	1166 sh	CH bend
1206 w	1184 w	1192 w	
	1240 w	1248 sh	CH rock
	1280 sh	1278 sh	
	1304 m	1300 sh	CC stretch
	1323 sh	1330 s	
	1362 w	1362 s	CC stretch
1391 m	1396 w	1400 m	CN stretch + NH stretch
1409 sh			CH rock
1427 vw		1431 w	
	1459 sh	1466 vw	
1488 vw	1477 sh	1488 sh	CC stretch
	1503 vs	1512 w	
1575 vw	1570 vs	1578 sh	CC stretch
1593 s	1594 s	1594 vs	CC stretch
1611 w	1618 s	1611 sh	NH2 bend
1673 m			C=O stretch

Abbreviations: w = weak m = medium; s = strong, v = very, sh = shoulder
Reprinted from J. Molec. Struct., 410–411, Iliescu T, Cinta S, Astilean S, Bratu I, pH influence on the Raman spectra of PP vitamin in silver sol, 193–196, copyright 1997, with permission from Elsevier

Usually, the aggregation of the colloid is a condition to observe SERS (Iliescu et al. 195). In this case, the aggregation was obtained by adding a small amount of $BaCl_2$ solution in the colloidal suspension, the first indication that the effect occurred being the change of color from yellow to green-blue (Iliescu et al. 1997).

From Fig. 6.8 one can observe that the substantial changes in the SERS spectra arise in the spectral range between 1300 and 1700 cm^{-1} as well as the reduction in intensity of the band 1027 cm^{-1} at low pH values. It is also worth noticing that the $C=O$ stretching band (1673 cm^{-1}) is not observed in all the SERS spectra. From the comparison of the SERS spectrum at pH = 5.5 with the Raman spectrum one

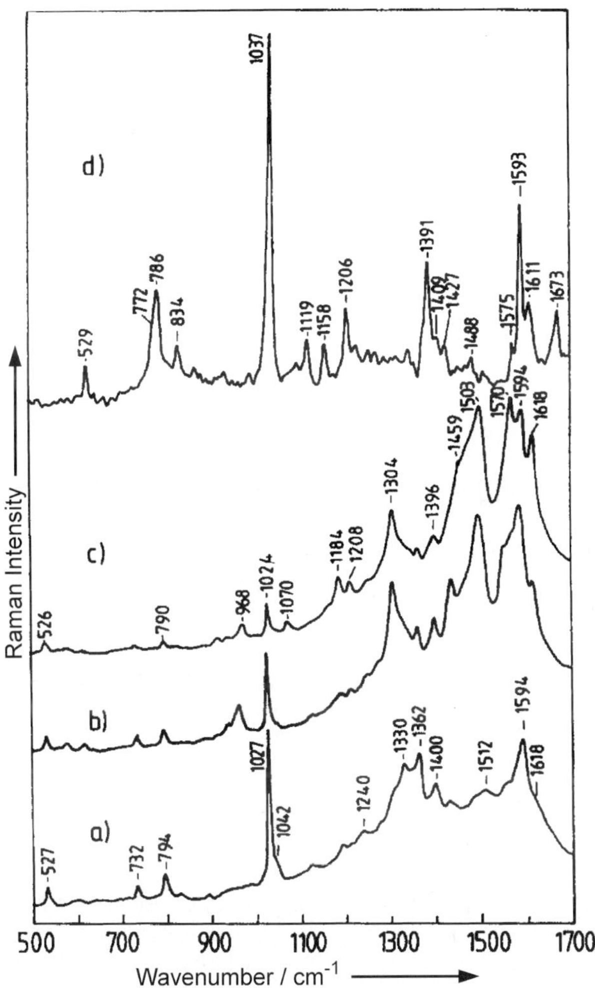

Fig. 6.8 SERS spectra of the PP vitamin at pH 5.5 (*a*), pH 3.5 (*b*), pH 2.5 (*c*), together with the Raman spectrum of solid state PP vitamin (*d*). Reprinted from J. Molec. Struct., 410–411, Iliescu T, Cinta S, Astilean S, Bratu I, pH influence on the Raman spectra of PP vitamin in silver sol, 193–196, copyright 1997, with permission from Elsevier

can readily see that the SERS spectrum correlates well with the Raman spectrum of the solid PP vitamin, which is in the nonprotonated form. This is an indication that at pH values equal with and above 5.5 the PP vitamin is adsorbed as the neutral form on the silver surface.

This supposition is confirmed by the presence of the band due to the ring stretching vibration at the same wavenumber value, as a dominant band, in the SERS spectrum at pH $= 5.5$ ($1594\,cm^{-1}$) and in the Raman spectrum of the solid PP vitamin at $1593\,cm^{-1}$. As the pH value of the solution is lowered, some new, intense peaks are developed at 1304, 1503, and $1570\,cm^{-1}$, characteristic for the PP vitamin protonated form. The presence of the band at $1594\,cm^{-1}$ indicates the existence of the neutral form even at pH $= 2.5$. Thus, at pH $<$ pK_a both protonated and neutral forms of the PP vitamin are adsorbed onto the silver surface (Iliescu et al. 1997). There are three possibilities for the PP vitamin to be bounded on the silver surface; one is via the lone pair electrons of the ring nitrogen, the second one is via the π electrons or nonbonding electrons of the amide group, and the third one is via the π ring electrons. In the amide group, there is a strong p-π conjugation represented by limited structures:

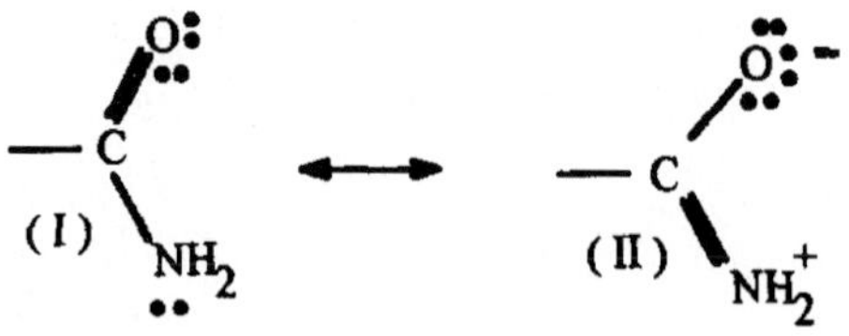

$$(\mathrm{I})\ \ \overset{\displaystyle O}{\underset{\displaystyle NH_2}{-C}} \quad\longleftrightarrow\quad (\mathrm{II})\ \ \overset{\displaystyle O^{-}}{\underset{\displaystyle NH_2^{+}}{-C}}$$

Probably, in the solution the II form is favored. This can explain the absence of the C=O stretching band in all SERS spectra. The peak position of the NH_2 bending is blue shifted by approximately $10\,cm^{-1}$, from $1611\,cm^{-1}$ in the Raman spectrum of the solid PP vitamin to $1618\,cm^{-1}$ in the SERS spectrum. This can indicate that the PP vitamin molecule is bonded by an amide group onto the silver surface via nonbonding electrons (Park et al. 1994). The noticeable enhancement of the NH_2 bending mode at low pH caused by either the proximity to the surface (electromagnetic enhancement) or the chemical effect (chemical enhancement) can imply that the amide group participates in the adsorption process.

The same change is observed for the band assigned to the CN stretching coupled with the NH bending mode, from $1391\,cm^{-1}$ in the Raman spectrum to $1400\,cm^{-1}$ in the SERS spectrum. The bonding of the PP vitamin via the amide group implies that the pyridine ring is probably in a vertical position on the silver surface. In SERS studies of pyridine the shift of the ring breathing vibration is taken as evidence for surface nitrogen interaction (Chang and Furtak 1982). One observes a shift of the band assigned to this vibrational mode from $1037\,cm^{-1}$ in the Raman spectrum of the PP vitamin to 1024 (1027) cm^{-1} in the SERS spectrum. This involves the participation of the ring nitrogen in the interaction with the silver surface, especially at large pH values, when this band is very intense. If the molecular plane lies flat on the surface, there exists the possibility of interaction between the π ring system and surface. It is known that a red shift by more than $10\,cm^{-1}$ of the ring vibrational modes as well as a substantial band broadening occurs as the benzene ring adsorbs onto the metal surface via the π system (Gao and Weaver 1985). A similar phenomenon can be expected to occur in the case of pyridine derivatives.

The fact that neither red shift nor band broadening occurs in the ring modes of the PP vitamin suggests that the interaction between the ring π system and the surface is not important. A similar situation was observed for nicotinic and isonicotinic acids (Park et al. 1994).

6.2.2 Conclusions

In the work presented herein SERS spectra of the PP vitamin adsorbed on colloidal silver nanoparticles surface were recorded and analyzed. It was established that the SERS spectra are preponderantly due to the neutral form of the PP vitamin at pH values equal with and above 5.5, and to the protonated form at acidic pH. It was also found that the protonated form is adsorbed via an amide group, while the neutral form is bound through the ring nitrogen atom.

References

Atkins PW (1987) Physikalische Chemie. VCH, Weinheim

Avram M, Mateescu GhD (1966) Applications of IR Spectra in organical chemistry. Editura Tehnica, Bucuresti

Breslow R (1958) On the mechanism of thiamine action. IV. Evidence from studies on model systems. J Am Chem Soc 80:3719–3726

Butler IS, Huang Y, Hadjiliadis N (1995) Pressure-tuning infrared spectra of the thiamine enzyme 'active aldehyde' intermediate 2-(α-hydroxycyclohexylmethyl)thiamine chloride (HCMT HCl) and its complex with zinc(II), Zn(HCMT)Cl3. Inorg Chim Acta 231:191–194

Chang RK, Furtak TE (1982) Surface-Enhanced Raman Scattering. Plenum Press, New York

Clarcke HT, Gurin S (1985) In: Bedeleanu DD, Manta I (eds) Biochimie Medicala si Farmaceutica, Editura Dacia, Cluj-Napoca

Creighton JA (1983) Surface Raman electromagnetic enhancement factors for molecules at the surface of small isolated metal spheres: The determination of adsorbate orientation from SERS relative intensities. Surf Sci 124:209–219

Dyda F, Furey W, Swaminnathan S, Sax M, Farrenkopf B, Jordan F (1993) Catalytic centers in the thiamin diphosphate dependent enzyme pyruvate decarboxylase at 2.4-Å resolution. Biochemistry 32:6165–6170

Friedemann R, Neef H (1988) Theoretical studies on the electronic and energetic properties of the aminopyrimidine part of thiamin diphosphate. Biochim Biophys Acta – Protein Structure and Molecular Enzymology 1385:245–250

Gao P, Weaver MJ (1985) Surface-enhanced Raman spectroscopy as a probe of adsorbate-surface bonding: Benzene and monosubstituted benzenes adsorbed at gold electrodes. J Phys Chem 89:5040–5046

Giese B, McNaughton D (2002a) Surface-enhanced Raman spectroscopic study of uracil. The influence of the surface substrate, surface potential, and pH. J Phys Chem B 106:1461–1470

Giese B, McNaughton D (2002b) Density functional theoretical (DFT) and surface-enhanced Raman spectroscopic study of guanine and its alkylated derivatives: Part 2: Surface-enhanced Raman scattering on silver surfaces. Phys Chem Chem Phys 4:5171–5182

Grecu I, Curea E (1980) Drugs identification. Editura Dacia, Cluj-Napoca

Iliescu T, Vlassa M, Caragiu M, Marian I, Astilean S (1995) Raman study of 9-methylacridine adsorbed on silver sol. Vib Spectrosc 8:451–456

Iliescu T, Cinta S, Astilean S, Bratu I (1997) pH influence on the Raman spectra of PP vitamin in silver sol. J Molec Struct 410–411:93–196

Jordan F, Xeng XP, Menon-Rudolph S, Barletta G, Annan N, Chung AC, Rios CB (1991) Observation and properties of the 2-α-carbanion (enamine) intermediate on pyruvate decarboxylase and in chemical models. In: Bisswanger H, Ullrich J (eds) Biochemistry and physiology of thiamin diphosphate enzymes. VCH, Weinheim

Kluger R (1987) Thiamin diphosphate: A mechanistic update on enzymic and nonenzymic catalysis of decarboxylation. Chem Rev 87:863–876

Leopold N, Cinta-Panzaru S, Baia M, Antonescu E, Cozar O, Kiefer W, Popp J (2005) Raman and surface-enhanced Raman study of thiamine at different pH values. Vib Spectrosc 39:169–176

Lindqvist Y, Schneider G, Ermler U, Sundström M (1992) Three-dimensional structure of transketolase, a thiamine diphosphate dependent enzyme, at 2.5 Å resolution. EMBO J 11:2373–2379

Louloudi M, Hadjiliadis N (1994) Structural aspects of thiamine, its derivatives and their metal complexes in relation to the enzymatic action of thiamine enzymes. Coor Chem Rev 135–136:429–468

Malandrinos G, Dodi K, Louloudi M, Hadjiliadis N (2000) On the mechanism of action of thiamin enzymes in the presence of bivalent metal ions. J Inorg Biochem 79:21–24

Moskovits M, Suh JS (1984) Surface selection rules for surface-enhanced Raman spectroscopy: calculations and application to the surface-enhanced Raman spectrum of phthalazine on silver. J Phys Chem 88:5526–5530

Muller YA, Schulz GE (1993) Structure and catalysis of the thiamine- and flavin-dependent enzyme pyruvate oxidase. Science 259:965–967

Park SM, Kim K, Kim MS (1994) Raman spectroscopy of isonicotinic acid adsorbed onto silver sol surface. J Molec Struct 328:169–178

Pletcher J, Sax M, Blank G, Wood M (1977) Stereochemistry of intermediates in thiamine catalysis. 2. Crystal structure of DL-2-(α-hydroxybenzyl)thiamine chloride hydrochloride trihydrate. J Am Chem Soc 99:1396–1403

Schellenberger A (1967) Struktur und Wirkungsweise des aktiven Zentrums der Hefe-Pyruvatdecarboxylase. Angew Chem 79:1050–1061

Schellenberger A, Schowen RL (1988) Thiamin Pyrophosphate Biochemistry. CRC Press, Boca Raton

Schowen BL, Schowen KB (1982) Solvent isotope effects on enzyme systems. Methods Enzymol 87C:551–606

Schrader B (1995) General Survey of Vibrational Spectroscopy. In: Schrader B (ed) Infrared and Raman Spectroscopy, Methods and Applications. VCH, Weinheim

Socrates G (2001) Infrared and Raman characteristic group frequencies: tables and charts. 3rd ed. Wiley, Chichester

Shin W, Oh DG, Chae CH, Yoon TS (1993) Conformational analyses of thiamin-related compounds. A stereochemical model for thiamin catalysis. J Am Chem Soc 115:12238–12250

Strekal ND, Gachko GA, Kivach LN, Maskevich SA (1992) SERS study of the complexes of thiamine derivatives with pyruvate. J Mol Struct 267:287–296

7 Other Molecules with Pharmacological Activity

7.1 2-Formylfuran Derivatives

2-Formylfuran (2FF) is a very important intermediate in organic synthesis. Some furan-based derivatives are muscarinic antagonists (Johansson et al. 1997) and show inhibitory activity towards cholesterol O-acyltransferase (Tanaka et al. 1998). A few 2-formylfuran derivatives (5-(4-fluor-phenyl)-2-formylfuran (5-(4FP)-FF) and 5-(4-brom-phenyl)-2-formylfuran (5-(4Br-P)-2FF) have been prepared and their bacteriostatic effects have been checked with good results. In order to know the action of potential drugs, such as the above-mentioned furan based derivatives, it is very important to see if the structure of the adsorbed species is the same as that of the free molecules. In these studies, a silver surface serves as an artificial biological interface (Dryhurst 1977). Prior to performing investigations on the above-mentioned derivatives, the interest was focused on the 2FF molecule.

7.1.1 2-Formylfuran

By the rotation of the COH group in the sample, two isomeric forms can be obtained. Several investigations have been performed on this molecular species up to now, and the conclusions drawn were quite different. Thus, the Raman and infrared data for the 2FF previously reported show intense doublets at 2800, 1670, 1420, and 1370 cm^{-1}. Furthermore, it was found that for solutions of 2FF in organic solvents, the intensity distribution of the doublets at 1670 and 1470 cm^{-1} is a function of concentration. By analyzing the infrared and Raman spectra of 2FF recorded at different temperatures, Allen and Bernstein (Allen and Bernstein 1959) concluded the existence of an equilibrium between the *cis-* and *trans*-isomer forms. They found no temperature dependence for the gas phase ($\Delta H \approx 0$), but for the liquid phase there was a small dependence corresponding to an energy difference of 4.185 kJ mol^{-1}. It was also showed by cryoscopy and infrared investigations that the carbonyl absorption around $1700\,cm^{-1}$ is not due to the molecular association of 2FF but to the existence of rotational *cis*- and *trans*-isomers. Karabatsos and Vane's nuclear magnetic resonance study (Karabatsos and Vane 1963) of coupling constants in a 2FF solution led to the result that the 2FF molecule exists completely in the *cis*-form. In contrast, Dahlqvist and Forsen (Dahlqvist and Forsen 1965), analyzing the nuclear magnetic resonance temperature dependence of 2FF dimethyl ether solution, found that the *trans*-form is more stable than the *cis*-form by $\Delta H = 4.39\,kJ\,mol^{-1}$. In a microwave and far-infrared study of the 2FF molecule it was found that in the vapor phase the molecule is planar and both rotamers co-exist, the *trans*-form being the most stable. Miller et al. (Miller et al. 1982) also found from far-infrared studies that the *trans*-isomer is more stable than the *cis*-isomer by $8.36\,kJ\,mol^{-1}$.

The lack of a consensus between the experimental data on 2FF reported in the literature requires supplementary theoretical and experimental studies of this molecular species. Therefore, an investigation of the two rotational isomers of 2FF from analytical (Raman spectroscopy) and theoretical (DFT calculations) points of view became necessary to be performed (Raman spectroscopy, surface-enhanced Raman spectroscopy and density functional theory studies of 2-formyl-furan, Iliescu T, Bolboaca M, Pacurariu R, Maniu D, Kiefer W, copyright 2003 John Wiley & Sons Limited. Reproduced with permission). In the first part of this analysis the experimental ΔH value, obtained from the plots of logarithmic relative intensities of different band pairs against reciprocal temperature, was determined and compared with the theoretical ΔH value obtained from DFT calculations, and in the second part SERS spectra of 2FF in silver colloid were also recorded and analyzed in order to determine the linkage of the two rotational isomers to the silver surface.

7.1.1.1 Vibrational Analysis

As it was already mentioned by rotation of the COH group in the sample, two configurations can be obtained. DFT calculations were performed at the BPW91/6–311+G* and B3LYP/6–311+G* theoretical levels to find the most stable configuration (Iliescu et al. 2003, Iliescu et al. 2002c). The optimized geometries of both conformations, calculated at the BPW91/6–311+G* level of theory, with the labeling of their atoms, are illustrated in Fig. 7.1.

Analytical harmonic vibrational modes were also calculated in order to ensure that the optimized structures correspond to minima on the potential energy surface. The total energy of the *trans*- and *cis*-isomers including zero point correction were found (Iliescu et al. 2003) to be −343.40906 and −343.40799 Hartree at the BPW91 theoretical level and −343.44103 and −343.43984 Hartree at the B3LYP theoretical level. Thus, at both of these levels of calculation, the *trans*-conformer was found to be more stable than the *cis*-conformer by $2.81\,kJ\,mol^{-1}$ (BPW91) and $3.12\,kJ\,mol^{-1}$ (B3LYP), less than the experimental values of 4.185 and $8.36\,kJ\,mol^{-1}$ obtained by Allen and Bernstein (Allen and Bernstein 1959)

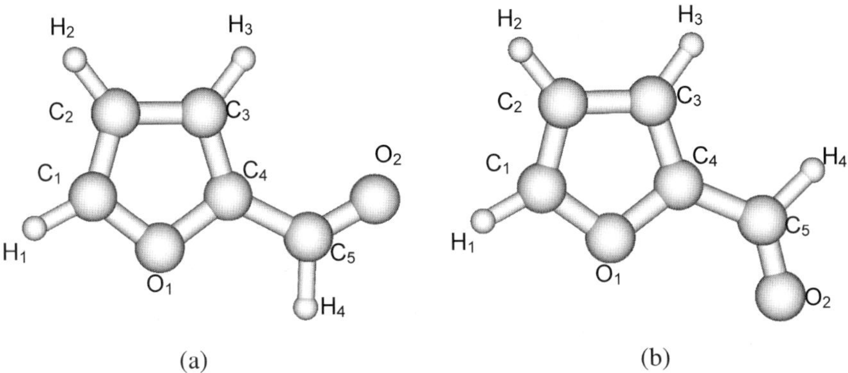

Fig. 7.1 Optimized geometries of the two isomers of 2FF: (**a**) *trans*-form and (**b**) *cis*-form (Raman spectroscopy, surface-enhanced Raman spectroscopy and density functional theory studies of 2-formylfuran, Iliescu T, Bolboaca M, Pacurariu R, Maniu D, Kiefer W, copyright 2003 John Wiley & Sons Limited. Reproduced with permission)

and Miller et al. (Miller et al. 1982), respectively. The enthalpy difference ΔH between the *cis*- and *trans*-isomers obtained also from theoretical calculations was found to be 2.99 kJ mol^{-1} (BPW91) and 3.20 kJ mol^{-1} (B3LYP), smaller than the experimental value obtained by Dahlqvist and Forsen (Dahlqvist and Forsen 1965) from nuclear magnetic resonance studies.

As already mentioned, the vibrational spectrum of 2FF was interpreted by Allen and Bernstein (Allen and Bernstein 1959) in terms of an equilibrium mixture of two rotational isomers with a planar configuration. They concluded that the *trans*-isomer is the most stable form in the liquid state of 2FF.

In order to eliminate the doubt concerning the differences between the experimental values of the energy and enthalpy of 2FF reported in the literature and the theoretical values obtained at the BPW91/6-311 + G* and B3LYP/6-311+G* levels, the temperature dependence of the Raman spectrum of the pure liquid was studied in the temperature range 274–365 K (Fig. 7.2).

From Fig. 7.2, one can see that the intensities of the bands at 1692, 1465, and 1371 cm^{-1} (the most representative bands were selected) increase with temperature and can be assigned to the less stable *cis*-isomer. The bands at 1673, 1477, and 1396 cm^{-1}, which are more intense at a low temperature, can be assigned to the most stable *trans*-isomer. Figure 7.3 shows the plots of logarithmic relative intensities of different band pairs (1673/1692, 1477/1465, 1396/1371) versus reciprocal temperature in the range 274–365 K. All pairs yield data points reasonably described by straight lines with almost identical slopes. From the slope of the plots, $\Delta H_{cis-trans} = 2.52 \pm 0.32$ kJ mol^{-1} was obtained.

The experimental ΔH value is fairly close to the theoretical ΔH values (2.99 kJ mol^{-1} (BPW91) and 3.20 kJ mol^{-1} (B3LYP)), the observed differences being largely due to the fact that the DFT calculations were performed for the gas phase, whereas the experimental results were obtained for the liquid phase.

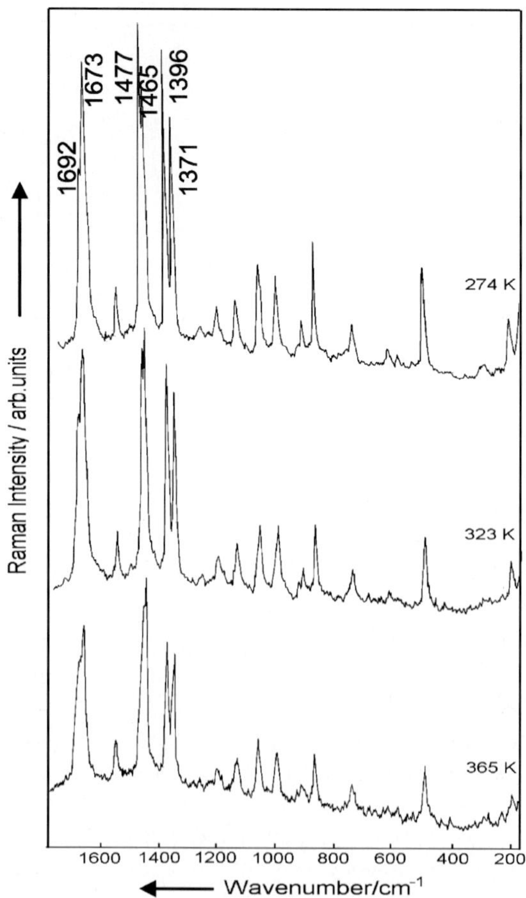

Fig. 7.2 Raman spectra of 2FF at different temperatures as indicated (Raman spectroscopy, surface-enhanced Raman spectroscopy and density functional theory studies of 2-formylfuran, Iliescu T, Bolboaca M, Pacurariu R, Maniu D, Kiefer W, copyright 2003 John Wiley & Sons Limited. Reproduced with permission)

Furthermore, the identification of the most stable form as the *trans-* or *cis-*conformer was obtained by comparing the experimental and calculated wavenumber shifts of the conformers. As can be observed from Table 7.1, the experimental wavenumber shifts ($\nu_{\text{least stable form}} - \nu_{\text{most stable form}}$) are correctly reproduced by the theoretical ($\nu_{\text{cis}} - \nu_{\text{trans}}$) values, thus clearly supporting the assignment of the most stable form to the *trans-*conformer (Iliescu et al. 2003, Iliescu et al. 2002c).

Another aspect that was necessary to be clarified, was the presence of the molecular association in the liquid state of 2FF. In this case, the monomer-dimer equilibrium can exist, the dimer being formed by hydrogen bond interaction with the carbonyl group. If the molecular association exists, the slope of the 1673/1692 pair assigned to the C=O stretching mode must be different from that of the other

Table 7.1 Observed and calculated *cis-trans* wavenumber shifts (cm^{-1}) of 2FF

ν_{exp} (cm^{-1})		$\Delta\nu_{exp}$ (cm^{-1})	$\nu_{theor.}^{a}$ (cm^{-1})		$\Delta\nu_{theor.}^{a}$ (cm^{-1})	$\nu_{theor.}^{b}$ (cm^{-1})		$\Delta\nu_{theor.}^{b}$ (cm^{-1})
cis	trans		cis	trans		cis	trans	
1692	1673	+19	1699	1690	+9	1697	1690	+7
1465	1477	−12	1456	1466	−10	1439	1450	−11
1371	1396	−25	1365	1394	−29	1349	1378	−29
2822	2859	−37	2808	2838	−30	2799	2824	−25

Abbreviations: [a] Obtained at the BPW91/6–311 + G* level, [b] obtained at the B3LYP/6–311 + G* level (Raman spectroscopy, surface-enhanced Raman spectroscopy and density functional theory studies of 2-formylfuran, Iliescu T, Bolboaca M, Pacurariu R, Maniu D, Kiefer W, copyright 2003 John Wiley & Sons Limited. Reproduced with permission)

pairs. This situation has been observed for other aldehydes such as 4-methylbenzaldehyde (Ribero-Claro et al. 1997). By inspecting Fig. 7.3, no difference between the slopes can be seen, and therefore the absence of the molecular association in the 2FF liquid state was assumed.

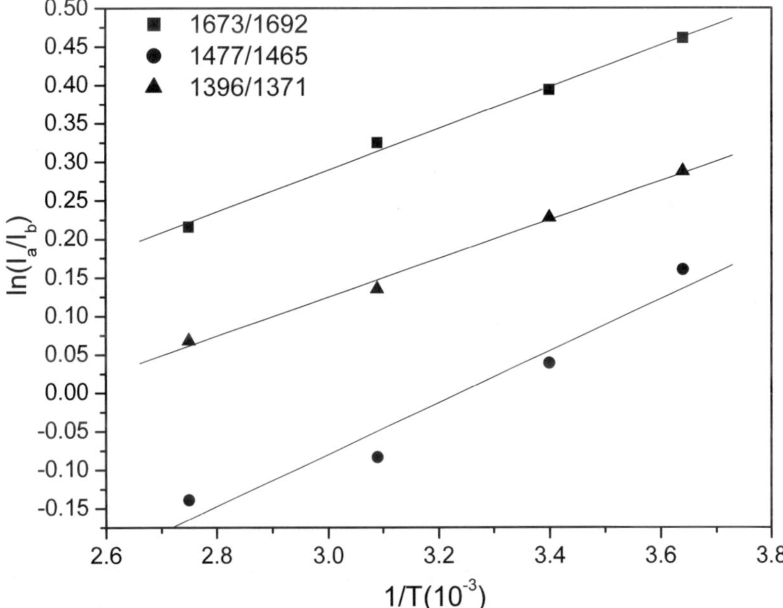

Fig. 7.3 Plots of logarithm of intensity ratio *vs* reciprocal temperature for the 1673/1692, 1477/1465, and 1396/1371 Raman band pairs (Raman spectroscopy, surface-enhanced Raman spectroscopy and density functional theory studies of 2-formylfuran, Iliescu T, Bolboaca M, Pacurariu R, Maniu D, Kiefer W, copyright 2003 John Wiley & Sons Limited. Reproduced with permission)

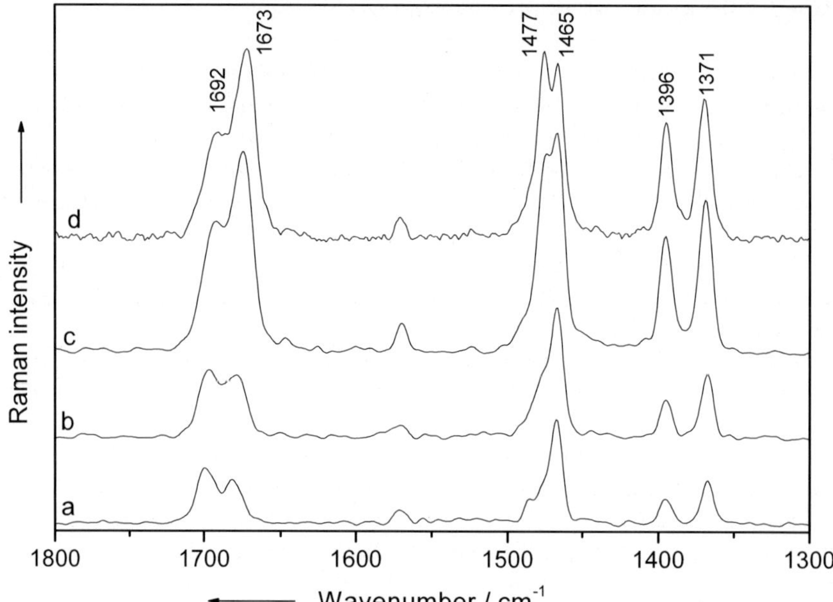

Fig. 7.4 Raman spectra of 2FF CCl$_4$ solution: (*a*) 10%, (*b*) 20%, (*c*) 60%, (*d*) 90% (Raman spectroscopy, surface-enhanced Raman spectroscopy and density functional theory studies of 2-formylfuran, Iliescu T, Bolboaca M, Pacurariu R, Maniu D, Kiefer W, copyright 2003 John Wiley & Sons Limited. Reproduced with permission)

A special behavior was observed for the pair of bands 1396/1371, which do not reverse their intensity on going from lower to higher temperatures (Fig. 7.2). Having in view that the slope of this pair is the same as that of the other pairs (see Fig. 7.3), one can suppose that the 1396/1371 bands are mainly due to the *trans*- and *cis*-form, respectively.

In order to explain the behavior of these bands, Raman spectra of different 2FF CCl$_4$ solutions were recorded and are presented in Fig. 7.4. As can be seen, the change in the relative intensities of this pair with dilution in CCl$_4$ is different from that corresponding to the other pairs (1673/1692, 1477/1465). While the intensity of the bands at 1673 and 1477 cm^{-1}, which are more intense at a low temperature and were assigned to the *trans*-isomer, decreases with dilution and the intensity of the bands at 1692 and 1465 cm^{-1} attributed to the *cis*-isomer increases, the relative intensity of the 1396/1371 pair remains constant for different 2FF concentrations. This behavior further supports the assumption that each of these bands is a superposition of the *cis*- and *trans*-isomer contribution (Iliescu et al. 2003, Iliescu et al. 2002c).

7.1.1.2 Adsorption on the Silver Surface

In order to determine the adsorption behavior of the two rotational isomers of 2FF on the silver surface, the SERS spectrum in a silver colloid at pH = 6 was recorded and is presented together with the FT-Raman spectrum in Fig. 7.5.

The observed Raman and SERS bands together with the vibrational assignment accomplished with the help of the results obtained from DFT calculations are summarized in Table 7.2.

The significant differences between FT-Raman and SERS spectra concerning relative intensities, bandwidths, and peak positions indicate an interaction between the metal and adsorbate.

If the molecules are physisorbed on the metal surface, the SERS spectrum is practically the same as that of the free molecules, small differences being observed only for the bandwidth (Bunding and Bell 1983). When the molecules are chemisorbed on the silver surface, there is an overlapping of the molecular and metal orbitals; the molecular structure of the adsorbate is modified (Creighton 1983) and in consequence the position and relative intensities of the SERS bands are dramatically changed. By comparing the SERS spectrum of 2FF with the conven-

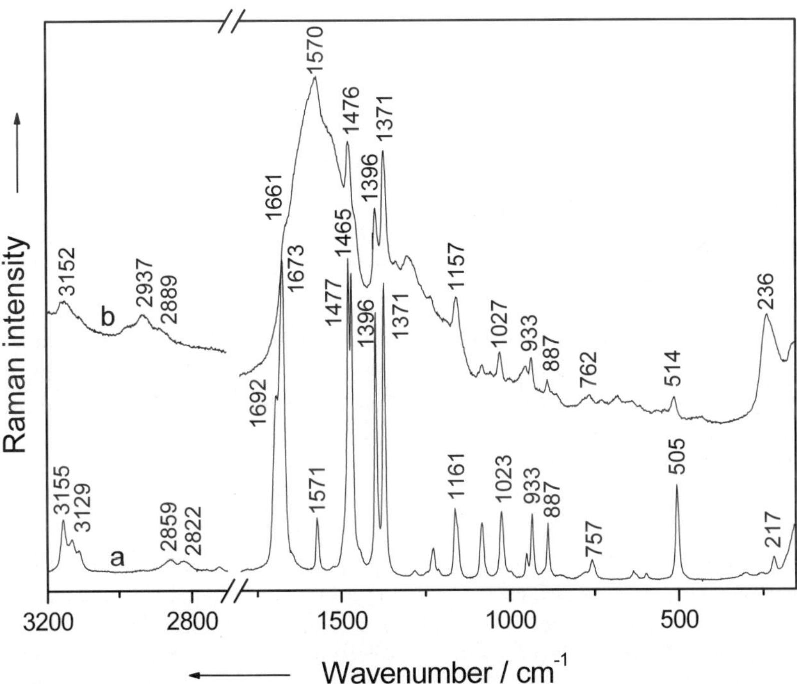

Fig. 7.5 FT-Raman (*a*) and SERS spectra (*b*) of 2FF in a silver colloid at the pH value of 6 (Raman spectroscopy, surface-enhanced Raman spectroscopy and density functional theory studies of 2-formylfuran, Iliescu T, Bolboaca M, Pacurariu R, Maniu D, Kiefer W, copyright 2003 John Wiley & Sons Limited. Reproduced with permission)

Table 7.2 FT-Raman and SERS wavenumbers (in cm^{-1}) at the pH value of 6 of 2FF and their assignments

Raman	Calc.[a] trans/cis	Calc.[b] trans/cis	SERS	Vibrational assignment
150 sh	199/197	199/197	–	$C_{3,4,5}$ bend
217 m	242/288	237/283	–	C_4C_5 wag
–	–	–	236 m	AgCl stretch
505 m	482/487	477/482	514 w	$O_1C_{4,5}$ bend
598 vw	586/584	583/583	–	Out-of-plane ring def
633 vw	620/635	615/630	–	
757 w	735/740	729/737	762 w	$O_2C_{5,4}$ bend
887 m	874/874	870/879	887 w	In-plane ring def
933 m	927/915	924/912	933 m	
951 w	963/961	968/965	951 w	CH wag (COH)
1023 m	1015/1021	999/1004	1027 w	CH bend (ring)
1061 m	1086/1080	1073/1066	1157 m	$C_1O_1C_4$ stretch
1226 w	1237/1257	1223/1240	1232 shw	$C_{4,5}$ stretch
1371 vs	1365/1394	1349/1351	1371 s	CH rock (COH) +
1396 vs	1405/1394	1392/1378	1396 m	$C_{2,3}$ stretch
1465 vs	1456/1466	1450/1439	1459 sh	
1477 vs	–	–	1476 s	Ring stretch
1571 m	1558/1550	1548/1541	1570	
1673 vs	1699/1690	1690/1697	1661 sh	CO stretch (COH)
1692 sh s	–	–	–	
2822 vw	2808/2838	2824/2799	2889 shw	CH stretch (COH)
2859 vw	–	–	2937 w	
3111 w	3179/3176	3126/3121	3108 sh	
3129 w	3198/3187	3141/3133	–	CH stretch (ring)
3155 m	3209/3209	3155/3155	3152 w	

Abbreviations: [a]Calculated with BPW91/6–311 + G[*], [b]Calculated with B3LYP/6–311 + G[*] and scaled by a 0.963 factor, w = weak, m = medium, s = strong, v = very, sh = shoulder, stretch = stretching, bend = bending, rock = rocking, twist = twisting, wag = wagging, ring = furan ring
(Raman spectroscopy, surface-enhanced Raman spectroscopy and density functional theory studies of 2-formylfuran, Iliescu T, Bolboaca M, Pacurariu R, Maniu D, Kiefer W, copyright 2003 John Wiley & Sons Limited. Reproduced with permission)

tional Raman spectrum (Fig. 7.5), shifts by 5–10 cm^{-1} of the peak positions can be observed. Therefore, one can conclude that 2FF molecules are chemisorbed on the silver surface (Iliescu et al. 2003, Iliescu et al. 2002c).

The background signal present in the SERS spectrum in the range 1600–1200 cm^{-1} is probably due to the photo- or thermal decomposition of 2FF, which forms a carbon layer on the silver surface (Iliescu et al. 1995). In fact, 2FF became colored after a very short time period and was distilled before each utilization.

The selective enhancement of the vibrational modes in the SERS spectrum was used as a probe for the determination of the adsorption mode of 2FF molecules on the silver surface. Thus, the band at $236\,cm^{-1}$ assigned to the AgCl$^-$ stretching mode presents an asymmetry at low wavenumbers. This asymmetry can be determined by the contribution of the AgO stretching vibration, although in the FT-Raman spectrum a very weak band is present at $217\,cm^{-1}$. These observations suggest (Sanchez-Cortes and García-Ramos 1990) that 2FF molecules are adsorbed on the metal surface through the oxygen atom. The enhancement of the band at $1157\,cm^{-1}$ assigned to the $C_1O_1C_4$ stretching vibration (see Fig. 7.1) in the SERS spectrum (Fig. 7.5) further demonstrates the adsorption of 2FF molecules on the silver surface through the ring oxygen atom (Iliescu et al. 2003, Iliescu et al. 2002c).

According to the electromagnetic surface selection rules (Creighton 1983, Iliescu et al. 1995, Sanchez-Cortes and García-Ramos 1990), a vibrational mode with its normal mode component perpendicular to the metal surface is likely to be more enhanced than a parallel one. In particular, the CH stretching vibrations have been reported to be a relatively unambiguous probe for adsorbate orientation (Moskovits and Suh 1984, Moskovits and Suh 1988, Gao et al. 1990). As can be observed from Fig. 7.5 and Table 7.2, the in-plane ring deformation modes at 887 and $933\,cm^{-1}$ and ring stretching modes at 1459, 1476, and $1570\,cm^{-1}$ are enhanced in the SERS spectrum in comparison to the corresponding bands from the conventional Raman spectrum. Moreover, the CH stretching bands appear distinct and enhanced in the SERS spectrum. All these features suggest that the 2FF molecules should assume a perpendicular or at least tilted orientation with respect to the silver surface (Iliescu et al. 2003, Iliescu et al. 2002c). The fact that the bandwidths are hardly affected by surface adsorption supports further the assumption that the flat orientation of 2FF molecules with respect to the surface is not likely (Oh et al. 1991).

From Fig. 7.5, one can see that bands due to the C=O stretching modes of the two isomers are not distinctly evidenced in the SERS spectrum. The enhancement and the strong shift of the carbonyl stretching mode present in the SERS spectrum of 2FF at $1661\,cm^{-1}$ suggest both the existence of a strong interaction between this group and the silver surface and the perpendicular orientation of this bond with respect to the metal surface. Moreover, the high intensity of the in-plane CH deformation vibration in the COH group that appears at $1371\,cm^{-1}$ in the SERS spectrum indicates the proximity of this group to the metal surface. This assumption is further supported by the shift of the CH stretching mode of this group observed in the SERS spectrum in comparison with the conventional Raman spectrum (see Table 7.2 and Fig. 7.5). By looking at the geometry of both rotational isomers and having in view that the molecule–substrate interaction is maintained both through the ring oxygen and oxygen atom of the substituent group, one can assume that the *cis*-isomer is mostly adsorbed on the silver surface. The enhancement of the ring stretching modes and CH deformation and stretching vibrations further supports this assumption and indicates the perpendicular or at least tilted orientation of the molecules on the metal surface (Iliescu et al. 2003, Iliescu et al. 2002c).

7.1.1.3 Conclusions

Raman spectroscopic investigations in combination with DFT calculations have been performed on 2FF. From DFT calculations performed at the BPW91/6–311 + G* and B3LYP/6–311 + G* theoretical levels it was found that the *trans*-isomer of 2FF is more stable than the *cis*-isomer by 2.81 and 3.12 kJ mol^{-1}, respectively. The theoretical ΔH values are very close to the experimental values determined from plots of logarithmic relative intensities of different band pairs of these isomers against reciprocal temperature. From the temperature dependence behavior of the carbonyl stretching mode, the absence of the molecular association of 2FF molecules in the liquid state was concluded. The SERS spectrum shows that 2FF molecules are chemisorbed on the silver surface through both the ring oxygen and the oxygen atom of the substituent group, the *cis*-form being preferred in the adsorbed state. The adsorbed molecules are oriented perpendicularly or at least tilted with respect to the silver surface.

7.1.2 5-(4-Fluor-phenyl)-2-formylfuran and 5-(4-Brom-phenyl)-2-formylfuran

After elucidating the structure and adsorption behavior of both rotational isomers of 2FF the attention was focused on the 5-(4-fluor-phenyl)-2-formylfuran (5-(4FP)-2FF) and 5-(4-brom-phenyl)-2-formylfuran (5-(4BrP)-2FF) derivatives analysis. In the next paragraphs, the rotational isomers of 5-(4FP)-2FF and 5-(4BrP)-2FF have been investigated by using infrared and FT-Raman spectroscopy in combination with DFT calculations (reprinted from Vib. Spectrosc., 29, Iliescu T, Irimie FD, Bolboaca M, Paisz Cs, Kiefer W, Vibrational spectroscopic investigations of 5-(4-fluor-phenyl)-furan-2-carbaldehyde, 235–239, copyright 2002, with permission from Elsevier). Having in mind that for understanding the action of potentially drugs, such as the above-mentioned furan based derivatives, it is very important to know if the structure of the adsorbed species is the same as that of the free molecules (Dryhurst 1977), SERS spectra of these compounds at low pH values have been also recorded and analyzed in order to elucidate the adsorption behavior of these molecules on colloidal silver particles (reprinted from Vib. Spectrosc., 29, Iliescu T, Irimie FD, Bolboaca M, Paisz Cs, Kiefer W, Surface-enhanced Raman spectroscopy of 5-(4-fluor-phenyl)-furan-2-carbaldehyde adsorbed on silver colloid, 251–255, copyright 2002, with permission from Elsevier).

7.1.2.1 Vibrational Analysis

By a rotation of the CHO group in the 5-(4FP)-2FF and 5-(4BrP)-2FF, two rotational isomers, the *cis*-form and *trans*-form, are obtained. The optimized geometries of these isomers, calculated at the BPW91/6–311 + G* level of theory with the labels of their atoms, are depicted in Fig. 7.6. The optimized structures of the isomers of both compounds are planar and belong to the C_s point group. Furthermore, the analytical harmonic vibrational modes have been calculated to ensure that the

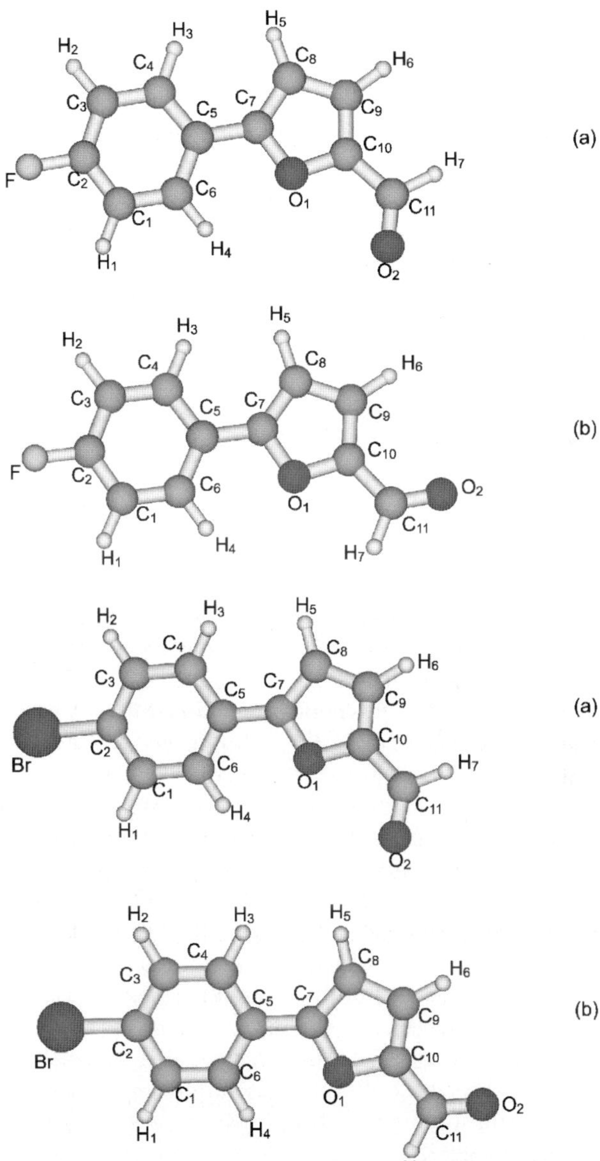

Fig. 7.6 Optimized geometries of the two isomers of 5-(4FP)-2FF and 5-(4BrP)-2FF: (a) *cis*-form isomer and (b) *trans*-form isomer. Reprinted from Vib. Spectrosc., 29, Iliescu T, Irimie FD, Bolboaca M, Paisz Cs, Kiefer W, Vibrational spectroscopic investigations of 5-(4-fluor-phenyl)-furan-2-carbaldehyde, 235–239, copyright 2002, with permission from Elsevier (5-(4FP)-2FF)

optimized structures correspond to minima on the potential energy surface. The total energy for the *cis*- and *trans*-form, including zero-point corrections, are found to be –673.44774 and –673.448048 Hartree, respectively for 5-(4FP)-2FF, while for 5-(4BrP)-2FF derivative they are –3147.9029 and –3147.9032 Hartree, respectively. Therefore, at this level of theory, for both compounds, the *trans*-form isomer was found to be more stable than the *cis*-form by 808.65 J mol^{-1} (5-(4FP)-2FF) (Iliescu et al. 2002a) and 795.87 kJ.mol^{-1} (5-(4BrP)-2FF) (Iliescu et al. 2003–2004). A similar situation was observed for the 2FF compound.

From the solid state sample, only strong fluorescence could be observed for the visible excitation wavelength. Therefore, near-infrared excitation was necessary to obtain Raman spectra. FT-Raman spectra of the polycrystalline samples are illustrated in Fig. 7.7 together with the calculated wavenumbers and intensities.

The observed bands in infrared and FT-Raman spectra of both compounds together with the calculated wavenumbers for both rotational isomers and the tentative assignment of the vibrational modes are summarized in Table 7.3. The assignment was accomplished mainly by comparison with related molecules (Dollish et al. 1973, Katritzky 1963, Mukherjee et al. 1997) and using the wavenumbers (unscaled values) and intensities as obtained by the BPW91 method. A strict comparison between the experimental and calculated wavenumbers and intensities is not possible in this case because the experimental data were obtained for a crystalline sample, whereas the theoretical calculations have been performed for the gas phase. Moreover, the calculated wavenumbers are obtained applying a harmonic approximation, whereas the experimental wavenumbers are of an anharmonical nature. Nevertheless, as revealed by Fig. 7.7 and Table 7.3, the quality of the quantum chemical results at the present theoretical level is sufficient for the assignment of the experimental data.

The crystalline 5-(4FP)-2FF and 5-(4BrP)-2FF contains both isomers, identified by the presence of two bands given by the C=O stretching vibrations, around 1661 and 1670 cm^{-1} in the Raman spectra and around 1675 and 1685 cm^{-1} in the infrared spectra. The assignment of these bands to the corresponding rotamer has been made with the help of theoretical calculations. Thus, the band at 1662 cm^{-1} (calc. 1679 cm^{-1}) is specific to the *trans*-form isomer, while the band at 1675 cm^{-1} (calc. 1682 cm^{-1}) corresponds to the *cis*-form isomer of 5-(4FP)-2FF. For the 5-(4BrP)-2FF compound the band at 1661 cm^{-1} (calc. 1680 cm^{-1}) is specific to the *trans*-form isomer, while the band at 1670 cm^{-1} (calc. 1689 cm^{-1}) corresponds to the *cis*-form isomer. Comparing the intensities of these two bands, one can infer that the *anti*-form is the preponderant rotamer in both solid state samples (Iliescu et al. 2002a).

Temperature dependent studies provide additional information concerning the most stable rotamer. The spectral region, corresponding to the C=O stretching vibration, in the infrared spectra of 5-(4FP)-2FF and 5-(4BrP)-2FF recorded at two different temperatures, 148 K and 298 K, respectively, is shown in Fig. 7.8. The intensity of the bands specific to the *trans*-form isomer (1670 cm^{-1} for 5-(4FP)-2FF and 1681 cm^{-1} for 5-(4BrP)-2FF), decreases as the temperature increases, and confirms the results obtained from theoretical calculations that this isomer is the more stable one (Iliescu et al. 2002a, Iliescu et al. 2003–2004).

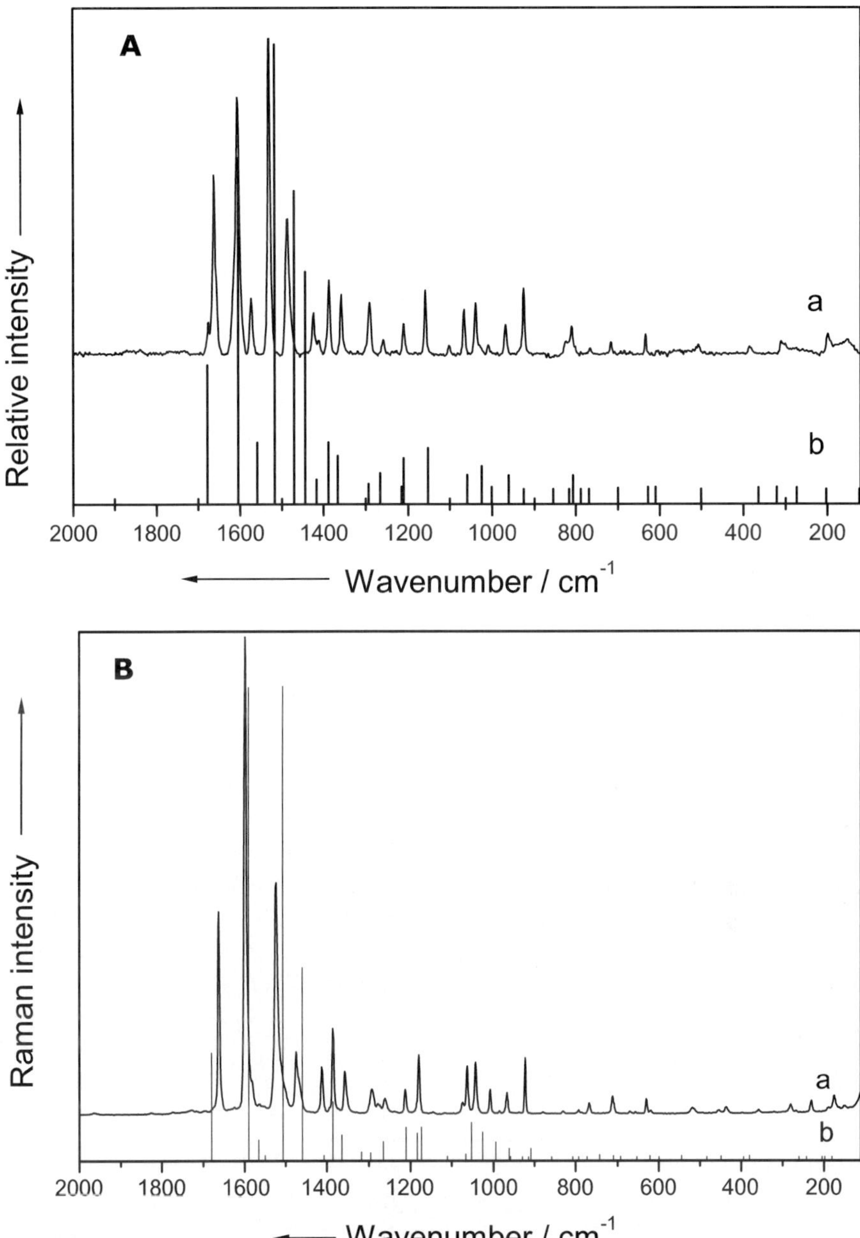

Fig. 7.7 FT-Raman spectrum (*a*) and the calculated Raman wavenumbers (*b*) of solid state 5-(4FP)-2FF (**A**) and 5-(4BrP)-2FF (**B**). Reprinted from Vib. Spectrosc., 29, Iliescu T, Irimie FD, Bolboaca M, Paisz Cs, Kiefer W, Vibrational spectroscopic investigations of 5-(4-fluor-phenyl)-furan-2-carbaldehyde, 235–239, copyright 2002, with permission from Elsevier (5-(4FP)-2FF)

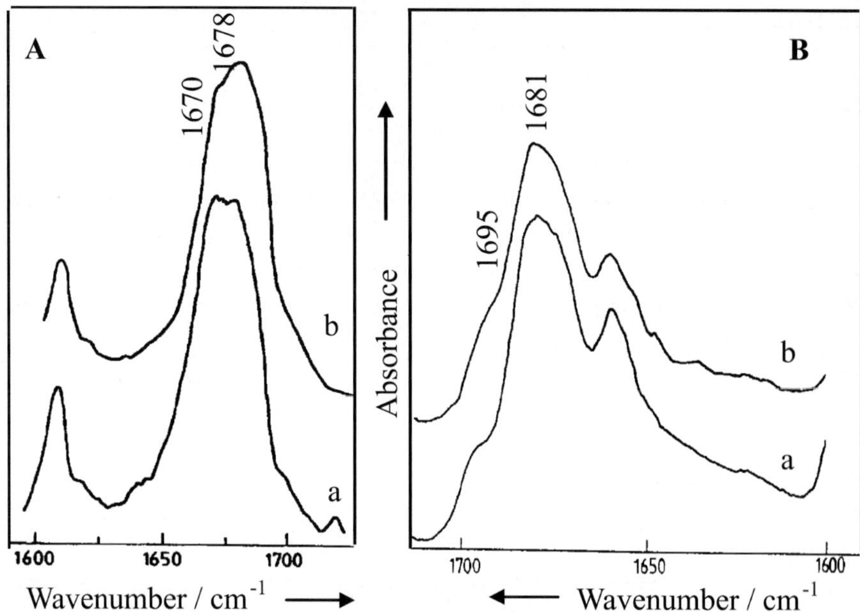

Fig. 7.8 The 1600–1700 cm^{-1} spectral region of the infrared spectra of 5-(4FP)-2FF (**A**) and 5-(4BrP)-2FF (**B**) recorded at different temperatures: (*a*) 148 K and (*b*) 298 K. Reprinted from Vib. Spectrosc., 29, Iliescu T, Irimie FD, Bolboaca M, Paisz Cs, Kiefer W, Vibrational spectroscopic investigations of 5-(4-fluor-phenyl)-furan-2-carbaldehyde, 235–239, copyright 2002, with permission from Elsevier (5-(4FP)-2FF)

As can be observed from Fig. 7.7 and Table 7.3, the bands given by the phenyl and furan ring stretching modes are present in the 1600–1400 cm^{-1} spectral region of the infrared and FT-Raman spectra of 5-(4FP)-2FF and 5-(4BrP)-2FF compounds. Most of the bands due to the in-plane CC and CH deformation vibrations of furan and phenyl rings can be observed in the 1200–1000 cm^{-1} spectral range. The 1000–400 cm^{-1} spectral region corresponds to out-of-plane ring and CH deformation vibrations (see Table 7.3).

Table 7.3 Experimental (infrared, FT-Raman) and calculated wavenumbers (cm^{-1}) (*trans/cis* forms) of 5-(4FP)-2FF and 5-(4BrP)-2FF compounds

5-(4FP)-2FF			5-(4BrP)-2FF			Vibrational
infrared	Raman	Calc.[a] trans/cis	infrared	Raman	Calc.[a] trans/cis	assignment
–	151 w	124/128	–	174 w	204/180	CO wag (COH)
–	199 w	203/177	–	229 w	242/221	$C_{10}C_{11}$ bend
–	–	–	–	280 w	260/260	CBr def + $C_4C_5C_7$ bend
–	300 vw	273/302	–	–	–	$C_4C_5C_7$ bend
–	310 w	321/329	359 vw	380/418		Ring 1[b] + ring 2[c] out of plane def

Table 7.3 (Continued)

| 5-(4FP)-2FF | | | 5-(4BrP)-2FF | | | Vibrational |
infrared	Raman	Calc.[a] trans/cis	infrared	Raman	Calc.[a] trans/cis	assignment
–	386 vw	365/371	–	–	–	CF bend + $C_{5,7,8}$ bend
510 m	507 vw	501/495	494 m	516 w	544/503	Ring 1 out of plane def
604 w	604 vw	610/625	–	–	–	$C_{1,2,3}$ bend
635 vw	634 w	628/653	628 w	630 m	622/621	$C_{2,1,6}$ bend + $C_{3,4,5}$ bend
668 m	716 w	700/694	712 m	711 m	709/704	Ring 1 out of plane def
765 m	764 vw	768/760	773 s	766 m	772/766	CH wag (ring 2)
787 m	802 sh	788/787	–	–	–	CH wag (ring 1)
804 w	809 w	807/806	793 vs	793 w	793/805	$C_{4,5,6}$ bend
833 m	823 sh	816/817	831 m	831 w	807/823	CH wag (ring 1)
879 vw	866 vw	855/838	880 vw	879 vw	858/850	CH twist (ring 2)
925 w	925 m	910/908	923 m	923 m	930/955	CH twist (ring 1)
967 m	968 w	961/951	967 s	967 m	962/960	$C_{9,10}O_1$ bend
1009 sh	1008 vw	1001/1001	1006 m	1007 m	994/993	$C_{1,2,3}$ bend + $C_{4,5,6}$ bend
1038 m	1039 m	1025/1032	1042 s	1041 m	1053/1058	CH bend (ring 2)
1066 vw	1066 m	1059/1058	–	–	–	$C_{5,7}O_1$ stretch
–	–	–	1073 m	1073 sh	1066/1070	CBr stretch + CH bend (ring 1)
1102 m	1102 vw	1101/1100	1170 w	1179 m	1184/1188	CH bend (ring 1) + $C_7O_1C_{10}$ stretch
1159 m	1158 m	1153/1153	–	–	–	
1212 sh	1211 mw	1211/1214	1212 w	1212 m	1211/1201	CH rock (ring 2)
1227 m	1220 vw	1216/1239	–	–	–	CF stretch + CH bend (ring 1)
1260 m	1258 w	1266/1282	1278 w	1278 m	1266/1248	$C_7O_1C_{10}$ stretch
1291 w	1290 m	1293/1294	1291 w	1293 m	1295/1292	CH rock (ring 1)
1357 w	1357 m	1366/1352	1357 w	1356 m	1364/1355	CH bend (COH) + $C_{5,7}$ stretch
1385 w	1387 m	1388/1388	1385 sh	1384 m	1385/1381	
1410 w	1411 sh	1417/1417	1411 m	1411 m	1407/1414	Ring 1 stretch + CH bend (COH)
1423 m	1424 w	1444/1450	–	–	–	
1487 vs	1485 s	1470/1473	1475 s	1474 m	1459/1469	Ring 2 stretch + $C_{5,7}$ + $C_{10,11}$ stretch
1528 v	1529 vs	1517/1516	1522 w	1522 s	1549/1553	Ring 2 stretch
1575 w	1573 m	1559/1554	–	1583 sh	1565/1569	Ring 1 stretch
1605 m	1606 vs	1604/1603	1597 m	1598 vs	1589/1596	
1670 sh	1662 s	1679/–	1681 vs	1661 s	1680/–	CO stretch (COH) trans-form
1678 vs	1675 sh	–/1682	1695 sh	1670 sh	–/1689	CO stretch (COH) cis-form
2853 w	2854 m	2834/2815	2848 w	2859 w	2830/2840	CH stretch (COH)

Table 7.3 (Continued)

| 5-(4FP)-2FF | | | 5-(4BrP)-2FF | | | Vibrational |
infrared	Raman	Calc.[a] trans/cis	infrared	Raman	Calc.[a] trans/cis	assignment
3099 w	3054 sh	3120/3119	3060 w	3058 sh	3117/3133	CH stretch (ring 1)
–	3080 m	3140/3140	–	3069 m	3129/3147	
3112 w	3094 sh	3183/3175	3112 w	3113 m	3140/3157	CH stretch (ring 2)
–	3116 w	3198/3191	–	3126 sh	3144/3160	

Abbreviations: [a]Calculated with BPW91/6–311+G*, ring 1[b] = phenyl ring, ring 2[c] = furan ring, w = weak, m = medium, s = strong, v = very, sh = shoulder, stretch = stretching, bend = bending, rock = rocking, twist = twisting, wag = wagging
Reprinted from Vib. Spectrosc., 29, Iliescu T, Irimie FD, Bolboaca M, Paisz Cs, Kiefer W, Vibrational spectroscopic investigations of 5-(4-fluor-phenyl)-furan-2-carbaldehyde, 235–239, copyright 2002, with permission from Elsevier (5-(4FP)-2FF)

7.1.2.2 Adsorption on the Silver Surface

The FT-Raman spectra of polycrystalline 5-(4FP)-2FF and 5-(4BrP)-2FF compounds together with their corresponding SERS spectra in a silver colloid at the pH value of 1 are illustrated in Fig. 7.9.

The assignment of the vibrational modes of the furan-based derivatives to the SERS bands at pH = 1 is summarized in Table 7.4.

In an alkaline environment, according to the *Cannizaro* reaction:

$$2[\text{R-CHO}] \quad \xrightarrow{[\text{OH}^-]} \quad [\text{R-CH}_2\text{OH}] + [\text{R-COOH}]$$

two species can be obtained from aldehydes. Furthermore, in the presence of a strong H donor group, like a hydroxyl group, a dimer can be obtained by an interaction with the carbonyl oxygen atom. Therefore, the SERS spectra of both furan-based derivatives have been analyzed only at an acidic pH value, due to the presence of many different species in an alkaline solution (Iliescu et al. 2002b, Iliescu et al. 2001, Iliescu et al. 2003–2004).

The significant differences between the FT-Raman and SERS spectra concerning the relative intensities, bandwidths, and peak positions indicate an interaction between the metal and adsorbate that causes a quite different derivative of the molecule's polarizability tensor. The spectra of physisorbed molecules are practically the same as those of the free molecules; small differences might be observed only for the bandwidths (Moskovits 1985, Vo-Dinh 1988). When the molecules are chemisorbed on a silver surface, an overlapping of the molecular and metal orbitals takes place causing significant changes in the position and relative intensities of the SERS bands (Campion and Kambhampati 1998, Lombardi et al. 1986). Comparing the SERS spectra of 5-(4FP)-2FF and 5-(4BrP)-2FF compounds to their corresponding conventional Raman spectra (see Fig. 7.9 and Table 7.4) shifts of the peaks position can be observed. Therefore, it can be concluded that both

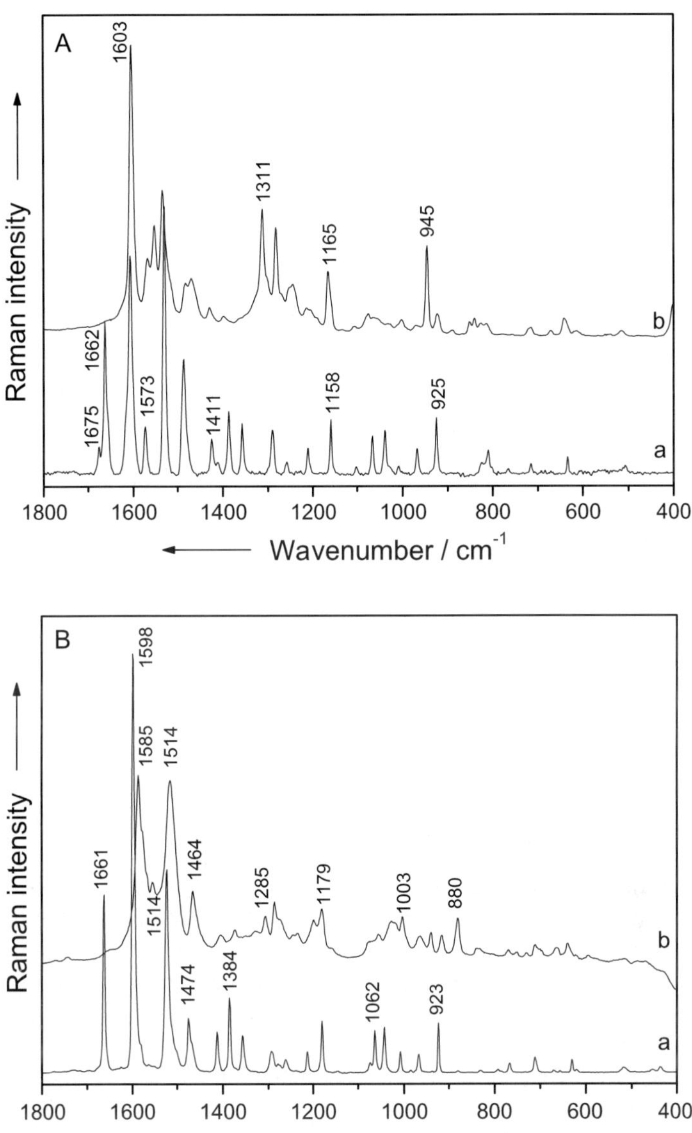

Fig. 7.9 FT-Raman (*a*) and SERS (*b*) spectra of 5-(4FP)-2FF (**A**) and 5-(4BrP)-2FF (**B**) in a silver colloid at the pH value of 1. Reprinted from Vib. Spectrosc., 29, Iliescu T, Irimie FD, Bolboaca M, Paisz Cs, Kiefer W, Surface-enhanced Raman spectroscopy of 5-(4-fluor-phenyl)-furan-2-carbaldehyde adsorbed on silver colloid, 251–255, copyright 2002, with permission from Elsevier (5-(4FP)-2FF)

Table 7.4 Assignment of the normal vibrational modes of 5-(4FP)-2FF and 5-(4BrP)-2FF to the SERS bands at the pH value of 1

5-(4FP)-2FF Raman	SERS pH 1	5-(4BrP)-2FF Raman	SERS pH 1	Vibrational assignment
199 w	190 vw	–	–	$C_{10}C_{11}$ def + AgO stretch
386 vw	401 w	–	–	CF bend + $C_{5,7,8}$ bend
507 vw	515 vw	–	–	Ring 1 out-of-plane def
604 vw	613 w	–	–	$C_{1,2,3}$ bend
634 w	641 w	630 m	639 w	$C_{2,1,6}$ bend + $C_{3,4,5}$ bend
716 w	671 vw	711 m	711 w	Ring 1 out-of-plane def
764 vw	756 vvw	766 m	770 w	CH wag (ring 2)
802 sh	779 vvw	793 w	–	CH wag (ring 1)
809 w	813 w	–	–	$C_{4,5,6}$ bend
823 sh	825 w	831 w	838 w	CH wag (ring 1)
866 vw	889 vw	879 vw	880 m	CH twist (ring 2)
925 m	923 w	923 m	916 w	CH twist (ring 1)
–	945 m	–	–	
968 w	969 vw	967 m	964 w	$C_{9,10}O_1$ bend
1008 vw	1002 vw	1007 m	1003 m	$C_{1,2,3}$ bend + $C_{4,5,6}$ bend
1066 m	1075 w	1041 m	1027 m	$C_{5,7}O_1$ stretch
1102 vw	1105 vw	–	–	CH bend (ring 1)
1158 m	1165 m	1179 m	1179 m	
1211 mw	1213 vw	1212 m	1198 m	CH rock (ring 2)
1220 vw	1244 m	–	–	CF stretch + CH bend (ring 1)
1258 w	1252 w	–	–	$C_7O_1C_{10}$ stretch
1290 m	1282 m	1293 m	1285 m	CH rock (ring 1)
–	1311 s	–	–	CH bend (COH) + $C_{5,7}$ stretch
1357 m	–	1356 m	1327 vw	
1387 m	1398 w	1384 m	1372 w	
1411 sh	–	1411 m	1404 w	Ring 1 stretch + CH bend (COH)
–	–	–	–	Ring 2 stretch
1424 w	1429 w	–	–	
–	1468 m	1474 m	1464 m	Ring 2 stretch + $C_{5,7}$ strech + $C_{10,11}$ stretch
1485 s	1482 w	–	–	
1529 vs	1533 m	1522 s	1514 s	Ring 2 stretch
–	1551 m	–	–	
1573 m	1567 m	1583 sh	1576 sh	Ring 1 stretch
1606 vs	1603 vs	1598 vs	1585 s	
1662 s	–	–	–	CO stretch (COH) trans-form
–	–	–	–	CO stretch (COH) cis-form
1675 sh	–	–	–	

Abbreviations: w = weak, m = medium, s = strong, v = very, sh = shoulder, stretch = stretching, bend = bending, rock = rocking, twist = twisting, wag = wagging. ring 1 = phenyl ring, ring 2 = furan ring

Reprinted from Vib. Spectrosc., 29, Iliescu T, Irimie FD, Bolboaca M, Paisz Cs, Kiefer W, Surface-enhanced Raman spectroscopy of 5-(4-fluor-phenyl)-furan-2-carbaldehyde adsorbed on silver colloid, 251–255, copyright 2002, with permission from Elsevier (5-(4FP)-2FF)

molecules are chemisorbed on the silver surface (Iliescu et al. 2002b, Iliescu et al. 2001, Iliescu et al. 2003–2004). In this case, it is difficult to differentiate between the contribution of the electromagnetic mechanism and charge-transfer effect, both contributing to the enhancement of the Raman signal. Additionally, some resonance Raman contribution to the total enhancement should not be excluded because the excitation wavelength of 514 nm falls on the wing of the absorption band of the 5-(FP)-2FF and 5-(4BrP)-2FF solutions.

From Fig. 7.9 and Table 7.4 one clearly sees that in the SERS spectrum at pH = 1 the strong band due to the C=O stretching vibration of the *anti*-form isomer is absent, while in the Raman spectrum this band appears around 1660 cm^{-1}. Mukherjee et al. (Mukherjee et al. 1997) also observed the absence of the carbonyl stretching mode in the SERS spectrum of the 2-isomer of formylpyridine, while the 3-isomer of formylpyridine showed an intense C=O stretching band. In the SERS spectrum of 2-formylthiophene, the band assigned to the C=O stretching vibration is more intense than in the spectrum of 3-formyltiophene (Mukherjee et al. 1997), and is determined by the dominance of the *cis*-form isomer in the surface adsorbed state. The absence or the extremely weak intensity of the carbonyl band in the SERS spectra of 4- and 2-acetylpyridine isomers on a silver electrode has been explained (Bunding and Bell 1996) considering the hydration of the C=O bond of the adsorbed molecules in the presence of water.

According to the electromagnetic surface selection rules (Moskovits and Suh 1984, Hallmark and Campion 1986, Moskovits and DiLella 1980) a vibrational mode with its normal component mode perpendicular to the metal surface is likely to become more enhanced than the parallel one. Gao and Weaver (Gao and Weaver 1985) observed a significant red shift (more than 10 cm^{-1}) of the ring stretching bands of the flat adsorbed aromatic molecules relative to the bulk spectra, due to the back-donation of electron density from the metal to the π^* antibonding orbital of the ring system.

In the SERS spectrum of 5-(FP)-2FF at the pH value of 1 (Fig. 7.9*a*) the bands at 1603 and 1533 cm^{-1} attributed to the ring stretching modes and the band at 1165 cm^{-1} assigned to the in-plane CH deformation vibration are more enhanced than other modes and red shifted by approximately 5 cm^{-1} in comparison to the bulk spectrum. At this pH value, the bands at 515 and 671 cm^{-1} due to out-of-plane deformation vibrations of the phenyl ring and the bands at 825, 889, and 923 cm^{-1} given by the CH wagging and CH twisting modes are only weakly enhanced. Therefore, one supposes that the molecular planes of adsorbed molecules are vertically orientated or less tilted with respect to the silver surface. The strong bands at 1311 and 945 cm^{-1} present in the SERS spectrum cannot be observed in the bulk spectrum and are probably due to a surface complex formed by adsorption (Iliescu et al. 2002b).

By analyzing the SERS spectrum of 5-(4Br-P)-2FF at the pH value of 1 (Fig. 7.9*b*) one can see that the bands at 1585, 1576, and 1514 cm^{-1} attributed to the ring stretching modes and the band at 1179 cm^{-1} assigned to the in-plane deformation vibration are more enhanced than other modes and shifted to lower wavenumber values by 8–13 cm^{-1} in comparison to those from the bulk spectrum.

At this pH value, the bands from the spectral range between 550 and 250 cm^{-1} given by the out-of-plane deformation vibrations of both rings are almost absent, while the bands at 916 and 838 cm^{-1} ascribed to the wagging and twisting vibrations of the CH groups (see Table 7.4) are only weakly enhanced. Therefore, one can suppose that the adsorbed 5-(4Br-P)-2FF molecules are vertical or less tilted relative to the silver surface.

If both rotamers are present in the solution and the molecular planes are orientated perpendicular or tilted on the metal surface, the C=O stretching mode of at least the *cis*-isomer should be present in the SERS spectrum, since this bond is approximately perpendicular to the silver surface. The absence of the bands corresponding to both isomers in the SERS spectrum could be a consequence of the hydration of the C=O bond according to the following reaction:

$$(R\text{-}C_6H_4\text{-}C_4H_2O)\text{-}CHO \quad \underset{\xleftarrow{\hspace{2cm}}}{\overset{H_2O}{\xrightarrow{\hspace{2cm}}}} \quad (R\text{-}C_6H_4\text{-}C_4H_2O)\text{-}CH(OH)_2$$

The red shift by 6 cm^{-1} and the enhancement of the band at 1252 cm^{-1} that involves a vibration of the furan ring oxygen reveal that the 5-(FP)-2FF molecules are chemisorbed on the silver surface via the lone pair electrons of the ring oxygen. Taking into account the weak enhancement of this band, one supposes that the interaction between the oxygen atom and the metal surface is not so strong. This assumption is further supported by the very weak intensity of the band at about 190 cm^{-1} assigned to the AgO stretching vibration (Iliescu et al. 2002b).

Similarly, from the SERS spectrum of 5-(BrP)-2FF molecules, the small shift and the enhancement of the band at 1041 cm^{-1} that involves the vibration of the furan ring oxygen indicates that the interaction between the oxygen atom and the metal surface via its lone pair electrons is not so strong (Iliescu et al. 2003–2004).

7.1.2.3 Conclusions

Infrared, FT-Raman spectroscopy, and SERS were applied to the vibrational characterization of 5-(FP)-2FF and 5-(BrP)-2FF derivatives. Theoretical calculations performed for both conformations of the samples revealed that the *trans*-form isomer is more stable than *cis*-form isomer by 808.65 J mol^{-1} in the case of the 5-(FP)-2FF compound and by 795.87 J.mol^{-1} for the 5-(BrP)-2FF derivative. The changes in the peaks' position and relative intensities observed in the SERS spectra compared to the FT-Raman spectra indicate the chemisorption of the 5-(FP)-2FF and 5-(BrP)-2FF molecules on the silver surface. The absence of the carbonyl band in the SERS spectrum was explained by the hydration of the C=O bond in the surface adsorption state. The molecules are adsorbed on colloidal silver particles via the nonbonding electrons of the ring oxygen and are perpendicularly orientated or at least tilted with respect to the silver surface.

7.2 Quinoline Derivatives

7.2.1 Isoquinoline

Isoquinoline (see Fig. 7.10) is a structural isomer of quinoline and represents the structural backbone in naturally occurring alkaloids, including papaverine and morphine. In contrast with quinoline, which is a hepatocarcinogen in mice and rats, and induces unscheduled DNA synthesis in primary cultures of rat hepatocytes, isoquinoline has not been shown to be genotoxic (LaVoie et al. 1983). Moreover, certain isoquinoline derivatives have shown anti-tumor properties (Dumaitre et al. 1997).

Isoquinoline has been identified as a component of coal tar, which behaves similarly to coal tar itself, and thus may contribute to the anti-psoriatic activity of coal tar (Foreman et al. 1985). Some isoquinoline derivatives have an anti-arrhythic and bradycardiac activity and are effective for the treatment of arrhythmia, myocardiac infarction or angina pectoris (Ishikawa et al. 1994).

The following paragraphs present an experimental and theoretical investigation of isoquinoline performed with the help of Raman spectroscopy and DFT calculations. Moreover, SERS spectra were also analyzed to elucidate the adsorption behavior of these molecules on colloidal silver particles. The adorption behavior of isoquinoline in acidic and alkaline environments was discussed in order to establish whether or not the molecule-substrate interactions are dependent on the pH of the solutions (Fourier transform Raman and surface-enhanced Raman spectroscopy of some quinoline derivatives, Bolboaca M, Kiefer W, Popp J, copyright 2002 John Wiley & Sons Limited. Reproduced with permission).

7.2.1.1 Vibrational Analysis

Isoquinoline is a colored liquid sample and present a large fluorescence at the visible excitation light; therefore, near-infrared excitation was necessary. The

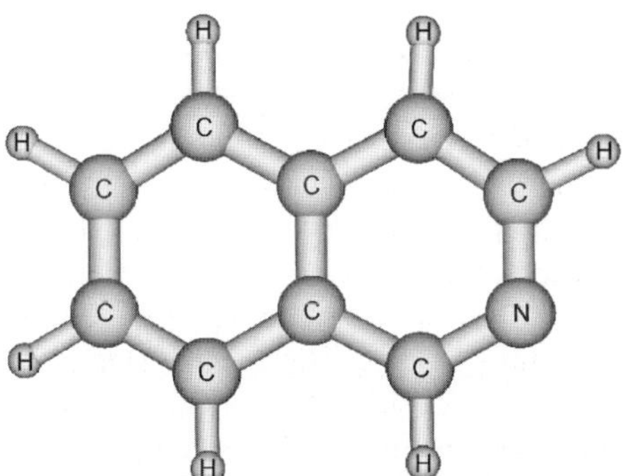

Fig. 7.10 Optimized structure of the isoquinoline molecule obtained at the BPW91/6–311 + G* theoretical level (Fourier transform Raman and surface-enhanced Raman spectroscopy of some quinoline derivatives, Bolboaca M, Kiefer W, Popp J, copyright 2002 John Wiley & Sons Limited. Reproduced with permission)

FT-Raman spectrum of the sample with the calculated unscaled Raman intensities are presented in Fig. 7.11.

The observed bands in the FT-Raman spectrum of isoquinoline with the tentative assignment of the vibrational modes are summarized in Table 7.5. The assignment was made with the help of results obtained from DFT calculations and the work of Wait and McNerney (Wait and McNerney 1970). Althrough the BPW91 method used for theoretical calculations do account for certain electronic

Table 7.5 Raman bands (experimental and calculated) of isoquinoline with their vibrational assignment

Raman Exp.	Calc.[a]	Vibrational assignment
381 w	367	Out-of-plane ring def
504 m	496	In-plane ring def
523 m	515	
540 sh	–	Out-of-plane ring def
639 w	632	
759 sh	766	In-plane ring def
783 s	771	
800 sh	791	
916 vw	925	CH wag
956 vw	928	In-plane ring def
1014 m	1017	Ring breathing (benzene)
1035 m	1037	Ring breathing (ring with N)
1140 w	1132	CH bend
1179 w	1180	
1258 w	1245	CH rock + CC stretch
1274 w	1260	
1329 m	1345	C_9C_{10} stretch
1383 vs	1371	CC stretch
1432 m	1429	CH rock
1461 m	1452	
1498 w	1498	CC stretch
1556 m	1561	CC stretch + CN stretch
1582 m	1579	CC stretch (ring with N)
1588 m	–	
1627 m	1620	CC stretch (benzene)
3053 s	3098	CH stretch

Abbreviations: [a] Calculated with BPW91/6–311 + G*, w = weak, m = medium, s = strong, sh = shoulder, stretch = stretching, bend = bending, wag = wagging, rock = rocking
(Fourier transform Raman and surface-enhanced Raman spectroscopy of some quinoline derivatives, Bolboaca M, Kiefer W, Popp J, copyright 2002 John Wiley & Sons Limited. Reproduced with permission)

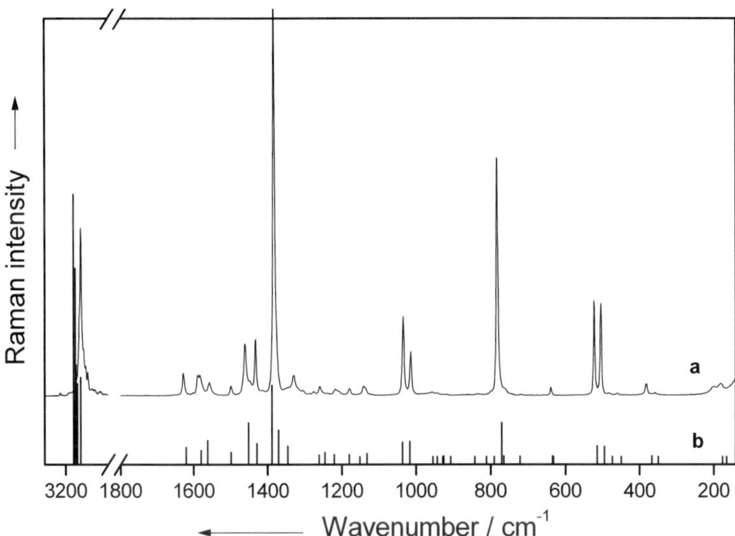

Fig. 7.11 FT-Raman spectrum (*a*) and the calculated Raman wavenumbers (*b*) of isoquinoline (Fourier transform Raman and surface-enhanced Raman spectroscopy of some quinoline derivatives, Bolboaca M, Kiefer W, Popp J, copyright 2002 John Wiley & Sons Limited. Reproduced with permission)

correlation effects, the differences between the experimental and calculated results arise from the presence of systematic errors due to the anharmonic effects and basis set deficiencies. However, as revealed by Fig. 7.11 and Table 7.5, the degree of agreement between the experimental and calculated wavenumbers was reasonably good.

The bands that dominate the FT-Raman spectrum (Fig. 7.11) are determined by the ring vibrations. Thus, the bands at 504 (calc. 496 cm^{-1}), 523 (calc. 515 m^{-1}) and 783 cm^{-1} (calc. 771 cm^{-1}) are due to the ring deformation vibrations, while the bands assigned to the ring breathing vibrations appear at 1014 (calc. 1017 cm^{-1}) and 1035 cm^{-1} (calc. 1037 cm^{-1}). The ring stretching modes give rise to the bands located in the 1600–1400 cm^{-1} spectral range. The other characteristic bands observed in the FT-Raman spectrum are due to the CH vibrations.

7.2.1.2 Adsorption on the Silver Surface

The low concentration ($\sim 10^{-4}$ M) required to obtain SERS spectra is a proof of the enhancement of the Raman signal. In order to understand the enhancement, UV-vis absorption spectra of a silver colloid before and after the addition of NaCl, and of the mixture of the activated colloid and isoquinoline were recorded and are illustrated in Fig. 7.12. The band at 408 nm in Fig. 7.12*a* is a characteristic of the plasma resonance adsorption for silver spheres in water. After the addition of NaCl this band is shifted to 420 nm. When isoquinoline is added to the silver

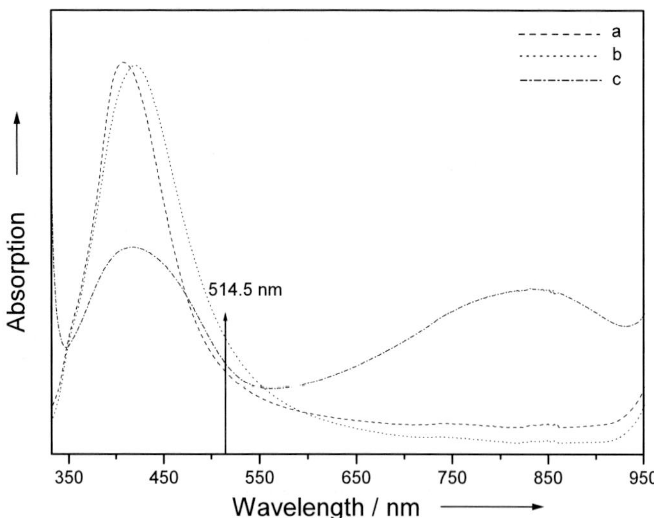

Fig. 7.12 Absorption spectra of (*a*) a pure silver colloid (*b*) 10^{-2} M NaCl added to a silver colloid with a volume ratio 1:10 (*c*) isoquinoline (overall concentration ~10^{-4} M) added to sample (*b*) (Fourier transform Raman and surface-enhanced Raman spectroscopy of some quinoline derivatives, Bolboaca M, Kiefer W, Popp J, copyright 2002 John Wiley & Sons Limited. Reproduced with permission)

colloidal suspension the absorption peak becomes weaker and broader and a new broad band around 850 nm appears, as can be clearly seen in Fig. 7.12*c*. This behavior is believed to be due to the formation of clusters of silver particles (Creighton et al. 1979). The main absorption maximum located near the wavelength of the incident light (514.5 nm) shows that the colloidal aggregate state is moderate and it is consistent with the experimental fact that the samples show very good SERS spectra in a silver colloid.

SERS spectra of isoquinoline adsorbed on colloidal silver nanoparticles at different pH values are illustrated in Fig. 7.13.

The assignment of the SERS bands observed at pH values of 1 and 14 to their corresponding Raman bands is presented in Table 7.6. By comparing the FT-Raman spectrum of isoquinoline with the SERS spectra differences in the position and relative intensities of the bands can be observed, this spectral behavior indicating a strong interaction between the metal and the adsorbate.

By comparing the SERS spectra of isoquinoline at pH values 1 and 14 with the FT-Raman spectrum a blue shift and an enhancement of the bands determined by the ring vibrations can be observed. Thus, the bands that appear in the 1600–1400 cm^{-1} range are shifted to higher wavenumbers in the SERS spectra by 3–15 cm^{-1}. The medium intense bands that appear in the FT-Raman spectrum at 504, 523, 783, 1014, and 1035 cm^{-1} are enhanced and shifted to 509, 525, 776, 1016, and 1036 cm^{-1} in the SERS spectrum recorded at pH = 1 and to 525, 787, 1018, and 1039 cm^{-1} in the SERS spectrum at pH = 14, respectively. The other bands present in the Raman spectrum of isoquinoline have the corresponding

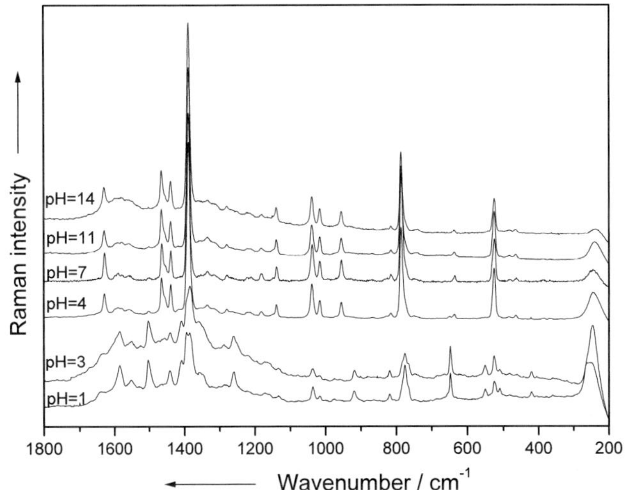

Fig. 7.13 SERS spectra of isoquinoline on silver colloids at different pH values as indicated (Fourier transform Raman and surface-enhanced Raman spectroscopy of some quinoline derivatives, Bolboaca M, Kiefer W, Popp J, copyright 2002 John Wiley & Sons Limited. Reproduced with permission)

bands shifted in the SERS spectra. Additionally, the occurrence of a new band at about $242\,\mathrm{cm^{-1}}$ in both SERS spectra, characteristic of an AgN stretching mode, suggests a chemical interaction between the silver substrate and the lone-pair electrons of the nitrogen. Therefore, one can assume that the isoquinoline molecules are chemisorbed on the silver surface through the nonbonding electrons of the nitrogen atom (Bolboaca et al. 2002).

By comparing the SERS spectra of isoquinoline at different pH values (Fig. 7.13) significant changes can be seen. Variation in SERS spectra with the change of the pH is usually attributed either to a reorientation of the adsorbed molecule with respect to the metal surface or to a change in its chemical nature (Rogers et al. 1984).

From Fig. 7.13 can be observed the different behavior of the adsorbed isoquinoline molecules on the silver surface in acidic and alkaline environments. In the SERS spectrum at the pH value of 14 a blue shift by $4\,\mathrm{cm^{-1}}$ of the bands from 1018 and $1039\,\mathrm{cm^{-1}}$, assigned to the ring breathing modes, can be seen. The in-plane ring deformation vibrations are enhanced and shifted to higher wavenumbers from 523, 783, and $956\,\mathrm{cm^{-1}}$ to 525, 787, and $958\,\mathrm{cm^{-1}}$, respectively, while the out-of-plane ring and CH deformation vibrations are only weakly enhanced or disappear in the SERS spectrum. The weak enhancement of the bands determined by the stretching vibration of the benzene ring could be a consequence of the relatively large distance between the ring and the metal surface. Taking into account the characteristics of the SERS spectra at pH values higher than 4 and having in mind the surface selection rules (Creighton 1983, Moskovits and DiLella 1980) one can assume that the adsorbed isoquinoline molecules are standing up on the surface (Bolboaca et al. 2002). Furthermore, the presence of the enhanced band at 3063 cm^{-1} due to the CH stretching vibration confirms this supposition (Moskovits and Suh 1988).

Table 7.6 Wavenumbers (cm^{-1}) and assignment of vibrational modes of the isoquinoline molecule to the SERS bands at pH values of 1 and 14

Raman	SERS		Vibrational assignment
	pH 1	pH 14	
–	247 s	237 w	AgN stretch
381 w	420 m	–	Out-of-plane ring dcf
504 m	509 m	–	In-plane ring def
523 m	525 m	525 m	
540 sh	550 m	–	Out-of-plane ring def
639 w	647 m	637 w	
759 sh	748 w	739 w	In-plane ring def
783 s	776 m	787 m	
800 sh	819 m	815 w	
916 vw	919 m	–	CH wag
956 vw	976 w	958 m	In-plane ring def
1014 m	1016 m	1018 m	Ring breathing (benzene)
1035 m	1036 m	1039 m	Ring breathing (ring with N)
1140 w	1134 w	1134 m	CH bend
1179 w	1172 m	1180 w	
1258 w	1259 m	1268 sh	CH rock + CC stretch
1274 w	1288 w	1279 w	
1329 m	1359	1354	C$_9$C$_{10}$ stretch
1383 vs	1386 s	1389 s	CC stretch
1432 m	1442 m	1439 m	CH rock
1461 m	1464 sh	1465 m	
1498 w	1503 m	1502 vw	CC stretch
1556 m	1552 m	1557 sh	CC stretch + CN stretch
1582 m	1585 m	1579 m	CC stretch (ring with N)
1588 m	–	1597 sh	
1627 m	1642 sh	1631 m	CC stretch (benzene)
3053 s	3066 m	3063 s	CH stretch

Abbreviations: w = weak, m = medium, s = strong, sh = shoulder, stretch = stretching, bend = bending, wag = wagging, rock = rocking
(Fourier transform Raman and surface-enhanced Raman spectroscopy of some quinoline derivatives, Bolboaca M, Kiefer W, Popp J, copyright 2002 John Wiley & Sons Limited. Reproduced with permission)

In spite of the differences between the SERS spectra at the pH values of 14 and 1, the bands due to the ring breathing modes are also shifted to higher waveumbers from 1014 and 1035 cm^{-1} to 1016 and 1036 cm^{-1}, respectively, in the SERS spectrum of isoquinoline at pH = 1. At this pH value the out-of-plane ring and CH deformation vibrations are enhanced and blue shifted from 381, 540, 639, and 916 cm^{-1} to 420, 550, 647, and 919 cm^{-1}, respectively, while the intensity of the band given by the CH stretching vibration is very weak. In this spectrum the benzene ring stretching vibrations are enhanced. Having in mind all these spectral features one can suppose that the changes between the SERS spectra of isoquinoline at different pH values reveal a reorientation of the adsorbed molecules with respect to the metal surface, the molecules being tilted on the surface at pH values lower than 4.

7.2.1.3 Conclusions

Experimental and theoretical investigations were performed on the isoquinoline molecule by using Raman spectroscopy and DFT calculations. From the analysis of the SERS spectra it was found that the behavior of the adsorbed molecules on colloidal silver nanoparticles is dependent on the pH conditions. The isoquinoline molecules are adsorbed via the nonbonding electrons of the nitrogen as neutral molecules, the variation in SERS spectra at different pH values being attributed to a change in orientation of adsorbed species.

7.2.2 Lepidine

Lepidine or 4-methylquinoline (see Fig. 7.14) is known as the most mutagenic form of the quinoline derivatives (Kato et al. 2000). However, some lepidine derivatives

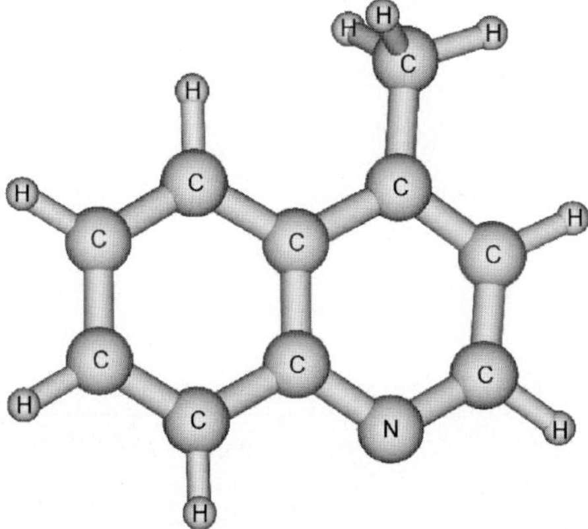

Fig. 7.14 Optimized structure of the lepidine molecule obtained at the BPW91/6–311 + G* theoretical level (Fourier transform Raman and surface-enhanced Raman spectroscopy of some quinoline derivatives, Bolboaca M, Kiefer W, Popp J, copyright 2002 John Wiley & Sons Limited. Reproduced with permission)

have shown to be antiprotozoal agents and afford improvement in means for the chemotherapy of leishmaniasis (Werbel and Steck 1987). Moreover, the most promising candidate for the prophylaxis of Chagas' disease, which is caused by the protozoan parasite *Trypanosoma cruzi*, is a hydrosoluble lepidine derivative, which was effective for the clearance of parasites from infected blood (Leite et al. 2006).

In this work, vibrational investigations of lepidine were performed by means of FT-Raman spectroscopy and DFT calculations. The analysis of the SERS spectra was also carried out in order to elucidate the adsorption behavior of these species on colloidal silver particles (Fourier transform Raman and surface-enhanced Raman spectroscopy of some quinoline derivatives, Bolboaca M, Kiefer W, Popp J, copyright 2002 John Wiley & Sons Limited. Reproduced with permission). The interest was also focused on discovering the adsorption behavior of lepidine on colloidal silver particles influenced by the pH values of the environment.

7.2.2.1. Vibrational Analysis

Lepidine is a colored liquid sample that exhibits under visible light excitation a large fluorescence. Therefore, for recording the Raman spectrum, near-infrared excitation was employed. Figure 7.15 shows the FT-Raman spectrum of lepidine with the calculated unscaled Raman intensities.

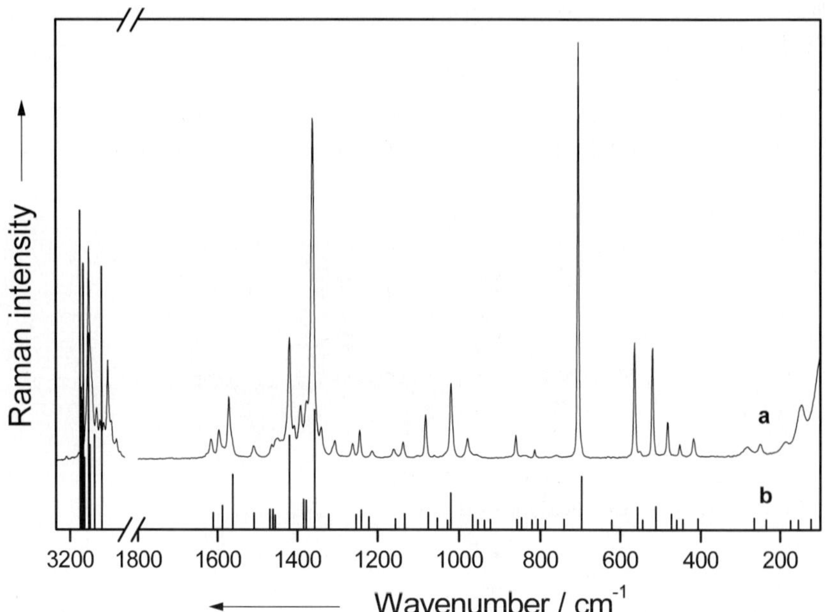

Fig. 7.15 FT-Raman spectrum (*a*) and the calculated Raman wavenumbers (*b*) of lepidine (Fourier transform Raman and surface-enhanced Raman spectroscopy of some quinoline derivatives, Bolboaca M, Kiefer W, Popp J, copyright 2002 John Wiley & Sons Limited. Reproduced with permission)

The observed bands in the FT-Raman spectrum together with the assignment of the vibrational modes, carried out with the help of results obtained from DFT calculations and the work of Wait and McNerney (Wait and McNerney 1970), are presented in Table 7.7. The differences evidenced between the experimental and

Table 7.7 Raman bands (experimental and calculated) of lepidine with their vibrational assignment

| Raman | | Vibrational |
Exp.	Calc.[a]	Assignment
418 m	407	Out-of-plane ring def
482 m	473	In-plane ring def
519 m	512	
551 sh	545	Out-of-plane ring def
564 m	558	
705 s	697	In-plane ring def
814 w	820	CH wag
860 m	857	
978 m	966	CH wag (CH_3)
1019 m	1020	Ring breathing (benzene)
1082 m	1077	Ring breathing (ring with N)
1137 w	1134	CH bend
1160 w	1157	
1244 m	1240	CH rock + CC stretch
1261 w	1253	
1306 m		C_9C_{10} stretch
1340 sh	1322	
1362 s	1357	CC stretch
1391 m	1385	CH rock
1419 m	1420	CH rock + CC stretch
1450 sh	1455	
1464 sh	–	
1507 w	1508	CC stretch
1565 sh	1561	CC stretch + CN stretch
1571 m	–	
1595 m	1587	CC stretch (ring with N)
1616 w	1611	CC stretch (benzene)
2922 m	2968	CH stretch (CH_3)
3063 s	3068	CH stretch

Abbreviations: [a] Calculated with BPW91/6–311 + G*, w = weak, m = medium, s = strong, sh = shoulder, stretch = stretching, bend = bending, wag = wagging, rock = rocking
(Fourier transform Raman and surface-enhanced Raman spectroscopy of some quinoline derivatives, Bolboaca M, Kiefer W, Popp J, copyright 2002 John Wiley & Sons Limited. Reproduced with permission)

calculated results mainly arise from the presence of systematic errors due to the anharmonic effects. However, as one can see from Fig. 7.15 and Table 7.7, the degree of agreement between the experimental and calculated wavenumbers was reasonably good.

The bands that dominate the FT-Raman spectrum (Fig. 7.15) are determined by the ring vibrations. Thus, the bands at 519 (calc. $512\,cm^{-1}$), 564 (calc. $564\,cm^{-1}$) and $705\,cm^{-1}$ (calc. $694\,cm^{-1}$) are due to the ring deformation vibrations, while those that appear at 1019 (calc. $1020\,cm^{-1}$) and $1082\,cm^{-1}$ (calc. $1077\,cm^{-1}$) are assigned to the ring breathing vibrations. The ring stretching modes give rise to the bands present in the $1600-1400\,cm^{-1}$ spectral range of the Raman spectrum of lepidine. The other characteristic bands observed in the FT-Raman spectrum are given by the CH vibrations.

7.2.2.2 Adsorption on the Silver Surface

For understanding the enhancement of the Raman signal, UV-vis absorption spectra of a silver colloid before and after the NaCl addition and also of the mixture of the activated colloid and lepidine were recorded and are shown in Fig. 7.16. The characteristic band of the plasma resonance adsorption for silver nanoparticles in water occurs at 408 nm (Fig. 7.16a). The NaCl addition causes a shift of this band to 420 nm. When lepidine is added to the activated colloid, the intensity of the

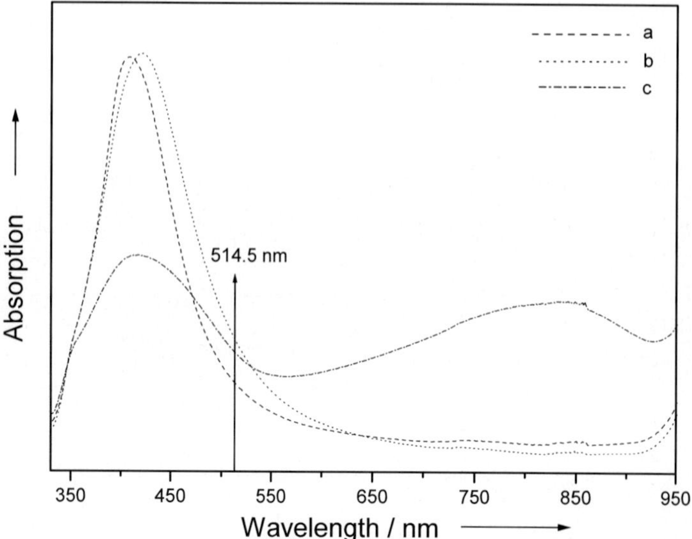

Fig. 7.16 Absorption spectra of (a) a pure silver colloid (b) 10^{-2} M NaCl added to a silver colloid with a volume ratio 1:10 (c) lepidine (overall concentration $\sim 10^{-4}$ M) added to sample (b) (Fourier transform Raman and surface-enhanced Raman spectroscopy of some quinoline derivatives, Bolboaca M, Kiefer W, Popp J, copyright 2002 John Wiley & Sons Limited. Reproduced with permission)

absorption peak decreases and it becomes broader, and a new broad band around 850 nm appears (Fig. 7.16*c*). This is an indication of the formation of clusters of silver particles and could explain the enhancement of the Raman signal of adsorbed lepidine molecule (Creighton et al. 1979).

SERS spectra of lepidine adsorbed on colloidal silver nanoparticles at different pH values are presented in Fig. 7.17. The assignment of the SERS bands observed at the pH values of 1 and 14 to their corresponding Raman bands is given in Table 7.8.

The differences observed between the SERS and Raman spectra regarding the position and relative intensities of some bands indicate a strong interaction between the metal and the adsorbate.

The comparison of the SERS spectra of lepidine at the pH values of 1 and 14 with the FT-Raman spectrum reveals different changes. Thus, in the SERS spectrum at the pH value of 1, the bands that appear in the $1600-1400\,\mathrm{cm}^{-1}$ range are shifted to lower wavenumbers by $2-6\,\mathrm{cm}^{-1}$, while in the SERS spectrum at pH = 14 these bands are shifted to higher wavenumbers by $1-5\,\mathrm{cm}^{-1}$. The medium intense bands present in the FT-Raman spectrum at 519, 564, 705, 1019, and $1082\,\mathrm{cm}^{-1}$ appear weakly enhanced at 524, 578, 707, 1025, and $1101\,\mathrm{cm}^{-1}$ in the SERS spectrum at the pH value of 1 and at 520, 579, 708, 1025, and $1108\,\mathrm{cm}^{-1}$ in the SERS spectrum at pH = 14, respectively. The band caused by the AgN stretching vibration can be observed only in the SERS spectrum at pH = 14, but significant changes in the relative intensities of some SERS bands compared to the nor-

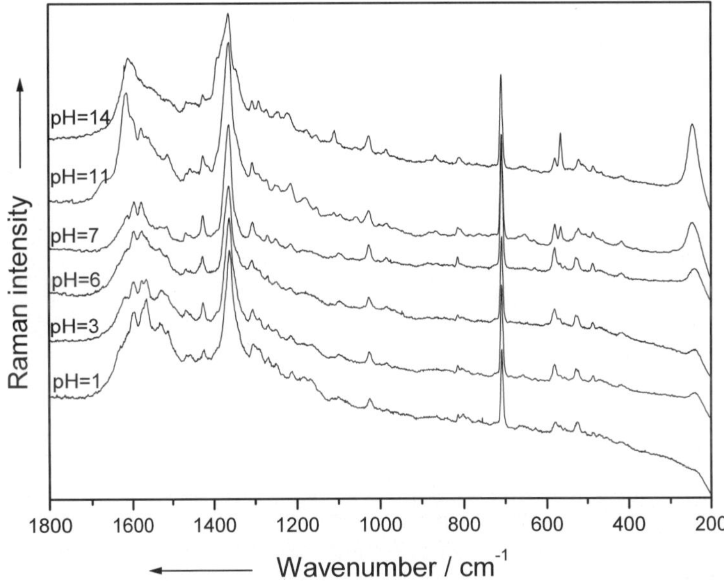

Fig. 7.17 SERS spectra of lepidine on silver colloids at different pH values as indicated (Fourier transform Raman and surface-enhanced Raman spectroscopy of some quinoline derivatives, Bolboaca M, Kiefer W, Popp J, copyright 2002 John Wiley & Sons Limited. Reproduced with permission)

Table 7.8 Wavenumbers (cm^{-1}) and assignment of vibrational modes of lepidine molecule to the SERS bands at pH values of 1 and 14

Raman	SERS pH 1	pH 14	Vibrational assignment
–	–	245 m	AgN stretch
418 m	417 vw	416 w	Out-of-plane ring def
482 m	486 w	485 w	In-plane ring def
519 m	524 m	520 m	
551 sh	567 sh	565 m	Out-of-plane ring def
564 m	578 m	579 sh	
705 s	707 s	708 s	In-plane ring def
–	789 w	–	NH def
814 w	813 w	808 w	CH wag
860 m	864 vw	866 w	
978 m	982 vw	984 w	CH wag (CH$_3$)
1019 m	1025 m	1025 m	Ring breathing (benzene)
1082 m	1101 w	1108 m	Ring breathing (ring with N)
1137 w	1168 m	1152 w	CH bend
1160 w	1183 m	1175 w	
1244 m	1250 w	1242 m	CH rock + CC stretch
1261 w	1268 w	1271 m	
1306 m	1304 m	1305 m	C$_9$C$_{10}$ stretch
1340 sh	1340 sh	1345 sh	
1362 s	1362 s	1363 s	CC stretch
1391 m	1391 sh	1389 sh	CH rock
1419 m	1423 m	1424 w	CH rock + CC stretch
1450 sh	1455 sh	1440 w	
1464 sh	1464 w	1465 sh	
1507 w	1509 sh	1512 sh	CC stretch
–	1530 m	–	NH def
1565 sh	1564 s	1553 sh	CC stretch + CNC stretch
1571 m	1570 sh	1578 sh	
1595 m	1595 s	1599 sh	CC stretch (ring with N)
1616 w	1609 sh	1610 s	CC stretch (benzene)
2922 m	2923 m	2921 m	CH stretch (CH$_3$)
3063 s	3065 mw	3068 ms	CH stretch

Abbreviations: w = weak, m = medium, s = strong, sh = shoulder, stretch = stretching, bend = bending, wag = wagging, rock = rocking

mal Raman spectrum can be observed in both spectra. The behavior of the adsorbed lepidine molecules denotes a partial chemisorption on the silver surface through the nonbonding electrons not only from the nitrogen atom but also from the ring (Bolboaca et al. 2002).

In the SERS spectra of lepidine at acidic pH (pH $\leq$ 6), two new bands located at 789 and 1530 cm^{-1} can be observed (Fig. 7.17). These bands were not present in the Raman spectrum and one supposes that they are due to the NH deformation vibrations (Levi et al. 1992). Taking into account their presence and the disappearance of the band at 245 cm^{-1} given by the AgN stretching vibration, one can assume that the lepidine molecules are protonated in an acidic environment (pK$_a$ = 5.67). The in-plane ring deformation vibrations are blue shifted from 482, 519 and 705 cm^{-1} to 486, 524 and 707 cm^{-1}, respectively and are enhanced, while the intensity of the band determined by the CH stretching vibration is weak in the SERS spectrum at these pH values. The bands at 1019 and 1082 cm^{-1} assigned in the Raman spectrum of lepidine to the ring breathing modes are shifted to higher wavenumbers to 1025 and 1101 cm^{-1}, respectively. The features of the SERS spectra at acidic pH values evidence the partial chemisorption of the protonated lepidine molecules via the π^* electrons of the ring, and their not perfectly parallel orientation to the silver surface. This supposition is confirmed by the red shift of the bands attributed to ring stretching vibrations by 1–7 cm^{-1} and the weak enhancement of the ring breathing modes (Gao and Weaver 1985, Park et al. 1994).

By comparing the SERS spectra of lepidine at alkaline pH values with the corresponding Raman spectrum, a blue shift by 3–20 cm^{-1} and an enhancement of the bands due to the ring breathing vibration and in-plane ring and CH deformation modes can be observed. The bands assigned to the CC and CH stretching vibrations are also enhanced at these pH values. The characteristics of the SERS spectra at an alkaline pH reveal a partial chemisorption of lepidine molecules on the silver surface via the nonbonding electrons of the nitrogen, the band at 247 cm^{-1} proving this assumption (Bolboaca et al. 2002). The weak enhancement of the bands assigned to the stretching vibrations of the benzene ring can be due to the relatively large ring-metal surface distance. Additionally, the enhancement of the band at 565 cm^{-1} due to the out-of-plane ring deformation vibration evidences that the molecules stand not perpendicularly but more likely tilted with respect to the silver surface.

7.2.2.3 Conclusions

In the work presented detailed FT-Raman spectroscopic investigations in combination with theoretical (DFT) calculations were performed on the lepidine molecules. SERS spectra analysis revealed that the adsorbed molecules behavior is strongly dependent on the pH conditions. At an acidic pH the lepidine molecules are preferentially adsorbed on the silver surface in the protonated form, via the π^* electrons of the ring and are not perfectly parallel with respect to the silver surface. At an alkaline pH the adsorbed molecules are tilted to the metal surface and bonded to it via the lone pair electrons of the nitrogen atom.

7.3 9-Phenylacridine

Acridine derivatives are one of the oldest classes of bioactive agents, which have been first developed as pigments and dyestuffs. Their pharmacological properties were evaluated during World War I mainly against bacterial infections, malaria, and other protozoa infections (Albert 1966, Wainwright 2001). The anticancer activity of acridines was first considered in the 1920s. Since then, a large number of acridine drugs have been tested as antitumor agents (Denny 2002, Demeunynck et al. 2001).

Acridine drugs display various chemical and biological properties but they share a common property of DNA intercalation (Denny 1990), because of the acridine chromophore presence that gives to the molecules a planar structure that allows them to bind DNA by stacking between base pairs. This intercalation property confers to the molecules a high affinity for DNA, which is generally considered as the biological target for acridine anticancer agents. However, it rapidly appeared that, while there was a good correlation between high affinity for DNA and cytotoxicity, intercalation capability was not sufficient for antitumor activities. It was therefore proposed that the antitumor properties of acridines were not only due to their mode of binding to DNA, but also to specific interaction with other nuclear receptors such as an enzyme (Liu 1989).

Having in mind these findings, the recent research is focused on designing new derivatives with anticancer potential and finding their structure-activity relationships for further use in clinical applications (Denny 2002).

From the multitude of acridine derivatives with pharmacological activity, 9-phenylacridine (9-PA) was selected and in the following paragraphs, a rather detailed analysis of its adsorption behavior on colloidal silver nanoparticles surface at different pH values is presented (Analyst, Iliescu T, Marian I, Misca R, Smarandache V, 119, 1994, 567–570. Reproduced by permission of The Royal Society of Chemistry).

7.3.1 Vibrational Analysis

Like acridine, 9-PA is soluble in an acidic media, forming its conjugate acid, 9-PAH$^+$. The dissociation constant of 9-PA in an aqueous solution determined by potentiometry was found to be $pK_a = 4.8$, lower than that of acridine, $pK_a = 5.6$ (Oh et al. 1991). The decrease of the pK_a value could be due to the phenyl radical that is attracting electrons from the acridine structure. Figure 7.18 shows the change of the 9-PA structure with the variation of the pH.

The Raman spectra of 9-PA in the solid state and solution at the pH value of 1.4 are illustrated in Fig. 7.19. By taking into account the pK_a value of 4.8, one can assume that only the 9-PAH$^+$ species are present in the solution.

The assignment of the bands appearing in the Raman spectra of 9-PA and 9-PAH$^+$ was carried out by using the spectral data for acridine and acridinium ion (Oh et al. 1991) and is summarized in Table 7.9.

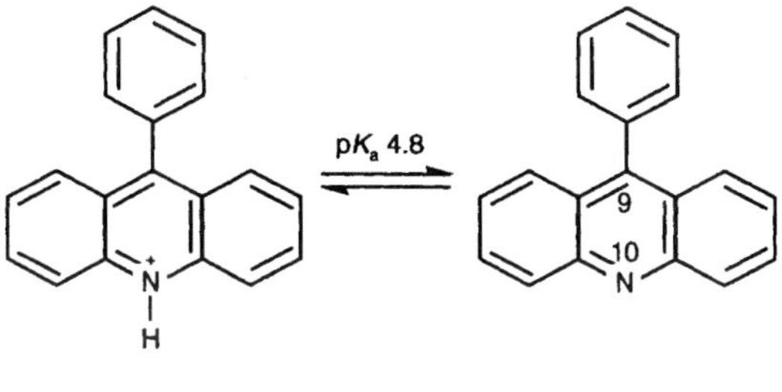

Fig. 7.18 Structure change of 9-PA with the changes of the pH (Analyst, Iliescu T, Marian I, Misca R, Smarandache V, 119, 1994, 567–570. Reproduced by permission of The Royal Society of Chemistry)

Table 7.9 Raman wavenumbers (in cm^{-1}) of 9-PA and 9-PAH$^+$ with their assignments

Raman Solid 9-PA	9-PAH$^+$	Vibrational assignment
145 m	165 w	Out-of-plane ring def
238 w	233 vw	Out-of-plane ring def
407 s	400 s	Out-of-plane ring def
435 s	425 m	
475 w	–	
487 m	485 w	Out-of-plane ring def
522 w	–	
572 vw	613 w	
613 w	–	
690 m	697 vw	CH wag
1000 m	999 vw	In-plane ring def
1017 w	1023 vw	
1165 w	1173 w	CH rock
1261 m	1273 s	
1287 w	1303 w	CH bend
1357 s	1371 s	
1410 vs	1411 vs	In-plane ring def
1475 w	1490 w	CC stretch
1510 vw	–	
1539 m	–	
1559 s	–	Ring stretch
1597 m	1588 m	
1605 m	1605 w	
–	1627 sh	NH def

Abbreviations: w = weak, m = medium, s = strong, v = very, sh = shoulder, stretch = stretching, bend = bending, rock = rocking, twist = twisting, wag = wagging
(Analyst, Iliescu T, Marian I, Misca R, Smarandache V, 119, 1994, 567–570. Reproduced by permission of The Royal Society of Chemistry)

Thus, the most intense bands that dominate the spectrum of 9-PA are due to the in-plane ring deformation vibrations and CH bending modes. By comparing the spectra of 9-PA and its ionic species, one can observe that a new band appears at $1627\,\mathrm{cm}^{-1}$, which is most probably due to the NH deformation vibration. Moreover, one can observe that the vibrational modes of 9-PA appear at higher wavenumber values when the molecules become protonated at the nitrogen atom.

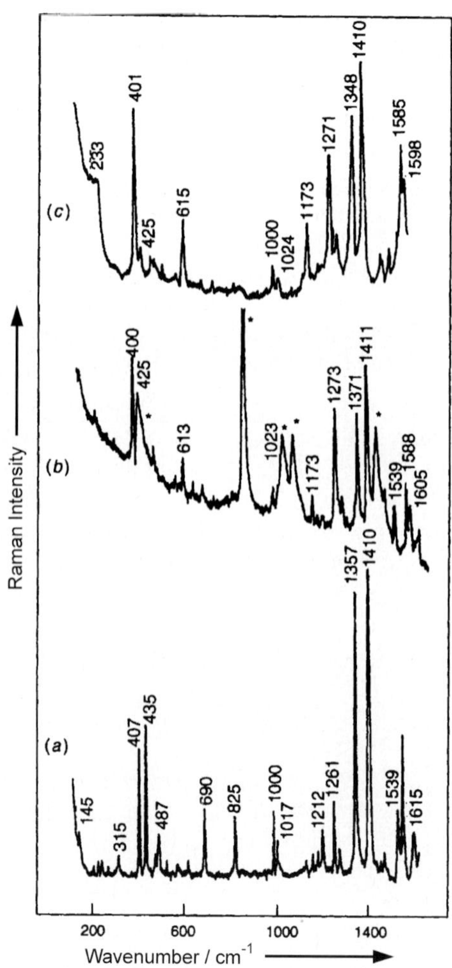

Fig. 7.19 Raman spectra of 9-PA in the solid state (*a*); ethanol-HCl-water solution at the pH value of 1.4 (*b*), and SERS spectrum of $5 \cdot 10^{-5}$ M 9-PA in activated silver colloid at the pH value of 1.4 (*c*). Asterisks indicate ethanol lines. (Analyst, Iliescu T, Marian I, Misca R, Smarandache V, 119, 1994, 567–570. Reproduced by permission of The Royal Society of Chemistry)

7.3.2 *Adsorption on the Silver Surface*

Prior to recording SERS spectra, the UV-vis absorption spectra of the silver colloids before and after adsorbate addition were recorded and are illustrated in Fig. 7.20. The freshly prepared colloid is yellow and shows a distinct absorption band at about 400 nm (Fig. 7.20*a*). The addition of 9-PA solution to a final concentration of $\sim 10^{-5}$ M caused a rapid color change to purple, which was further intensified by the KCl (10^{-3} M) addition. In the absorption spectrum of a silver colloid with 9-PA (Fig. 7.20*b*), the band at 400 nm is shifted to a longer wavelength and a new band is developed at about 480 nm. These results suggest the colloid aggregation, which was associated with increased surface-enhanced Raman scattering by an electromagnetic mechanism (Creighton 1986).

SERS spectra of 9-PA recorded at different pH values in the KCl presence are shown in Fig. 7.21. The chloride ions addition was necessary because it induces an enhancement of the Raman signal (Chang and Furtak 1982) and prevents photoreactions on the silver surface (Oh et al. 1991).

The assignment of the SERS bands observed at pH values of 1.4. and 10.7 to their corresponding Raman bands is given in Table 7.10.

As was already mentioned, the Raman spectrum of the 9-PA solution at the pH value of 1.4 can be almost entirely attributed to the 9-PAH$^+$ ion. By comparing the SERS spectrum recorded at pH = 1.4 (Fig. 7.19*c*) with the Raman spectrum of 9-PA solution at the same pH (Fig. 7.19*b*) one can observe that they look similar. This confirms that at this pH value, only the protonated form is adsorbed on the silver surface (Iliescu et al. 1994).

On the other hand, the comparison of the SERS spectrum recorded at the pH value of 10.7 with the Raman spectrum of the polycrystalline sample reveals only small

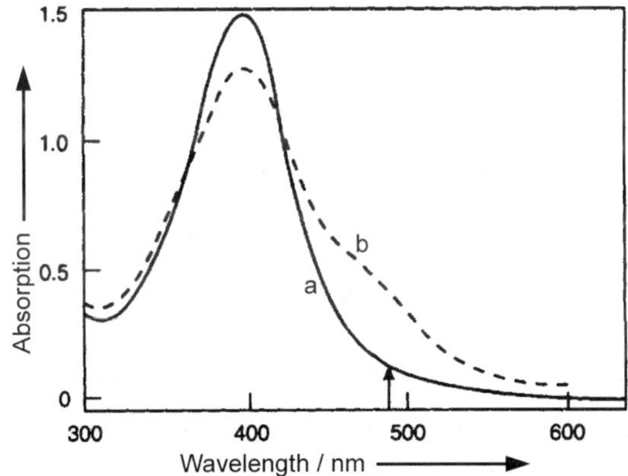

Fig. 7.20 UV-vis absorption spectra of a silver colloid before (*a*) and after (*b*) the 9-PA addition. The arrow shows the position of the exciting laser radiation at 488 nm (Analyst, Iliescu T, Marian I, Misca R, Smarandache V, 119, 1994, 567–570. Reproduced by permission of The Royal Society of Chemistry)

shifts of the SERS bands. Thus, one can assume that at this pH value 9-PA adsorbed on the silver colloid surface exists in the unprotonated form (Iliescu et al. 1994).

From the SERS spectra recorded at the pH values of 4.5, 6.6, 10.7, and 12 (Fig. 7.21) one can see that with the pH increase the intensity of the bands at 1266, 1365, and 1570 cm^{-1} increases, while the intensity of the bands at 1277, 1374, and 1590 cm^{-1} decreases, and at the pH value of 10.7 these are not present in the spectrum anymore. Therefore, the first set of bands can be assigned to the 9-PA neutral form and the latter bands to the protonated form. In fact, the band that appears around 1370 cm^{-1} reveals an overlap of two bands at 1365 and 1374 cm^{-1}, which changes their relative intensities with increasing pH values.

Several cases have been reported in which the pK$_a$ values of acids are modified by surface adsorption (Sun et al. 1985). This can explain why both neutral and

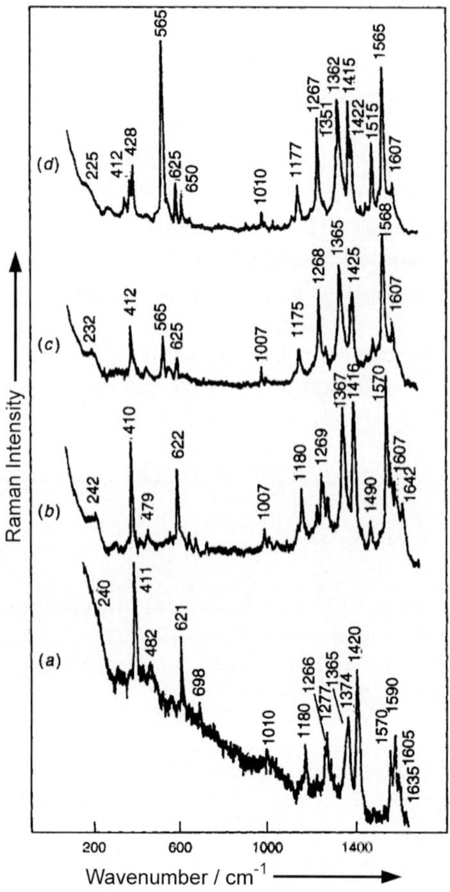

Fig. 7.21 SERS spectra of 9-PA at different pH values: (*a*) pH 4.5, (*b*) pH 6.6, (*c*) pH 10.7 and (*d*) pH 12 (Analyst, Iliescu T, Marian I, Misca R, Smarandache V, 119, 1994, 567–570. Reproduced by permission of The Royal Society of Chemistry)

Table 7.10 Raman and SERS wavenumbers (in cm^{-1}) of 9-PA at different pH values and their assignments

Raman		SERS		Vibrational
solid 9-PA	9-PAH$^+$	pH 1.4	pH 10.7	assignment
145 m	165 w	–	–	Out-of-plane ring def
238 w	233 vw	233 w	235 w	Out-of-plane ring def + AgCl stretch
407 s	400 s	401 vs	412 m	Out-of-plane ring def
435 s	425 m	425 w	428 w	
475 w	–	470 w	–	
487 m	485 w	488 w	480 vw	Out-of-plane ring def
522 w	–	520 w	–	
572 vw	–	–	565 m	
613 w	613 w	615 m	625 w	
690 m	697 vw	673 vw	–	CH wag
1000 m	999 vw	1000 w	1007 vw	In-plane ring def
1017 w	1023 vw	1024 vw	–	
1165 w	1173 w	1173 m	1175 m	CH rock
1261 m	1273 s	1271 s	1268 s	
1287 w	1303 w	1303 m	1295 w	CH bend
1357 s	1371 s	1368 vs	1365 s	
1410 vs	1411 vs	1410 vs	1410 m	In-plane ring def
–	–	–	1425 s	
1475 w	1490 w	1488 w	1485 vw	CC stretch
1510 vw	–	–	1516 w	
1539 m	–	1528 w	–	
1559 s	–	–	1568 vs	Ring stretch
1597 m	1588 m	1585 m		
1605 m	1605 w	1598 m	1607 m	
–	1627 sh	1635 m	–	NH def

Abbreviations: w = weak, m = medium, s = strong, v = very, sh = shoulder, stretch = stretching, bend = bending, rock = rocking, twist = twisting, wag = wagging
(Analyst, Iliescu T, Marian I, Misca R, Smarandache V, 119, 1994, 567–570. Reproduced by permission of The Royal Society of Chemistry)

protonated forms of 9-PA are adsorbed over a wide range of pH (4–9) values. Thus, in the SERS spectra at the pH values of 4.5 and 6.6, both 9-PA species are adsorbed on the silver surface.

9-PA may bind to the surface via its ring π system or via nitrogen lone pair electrons. It is known that the frequencies of the ring stretching vibrations decrease by more than $10\,cm^{-1}$ and their bandwidths increase substantially when molecules adsorb on the metal surface via their π system (Oh et al. 1991). In the case of the 9-PA molecule, the bands due to the ring stretching vibrations are shifted to higher wavenumbers by 6–9 cm^{-1} in the SERS spectra and the bandwidths are hardly affected by adsorption. Therefore, one can assume that surface-ring π orbital overlap does not occur significantly when 9-PA adsorbs on the silver surface.

The shift of most of the SERS bands in comparison with the corresponding Raman bands indicates that the 9-PA molecule is bound to the silver surface via nitrogen lone pair electrons.

It is known (Oh et al. 1991) that a charged cation is not efficiently adsorbed on the colloidal silver surface. Considering that Cl$^-$ is needed to obtain the SERS spectrum of 9-PAH$^+$, one can infer that the 9-PAH$^+$ ion is paired with the surface chloride and this ion pair gives rise to the bands from the SERS spectra recorded at pH = 1.4 and partially to those from the spectra obtained at higher pH values (4.5 and 6.6). A similar mechanism has been reported for acridinium ion (Oh et al. 1991).

By inspecting the Raman spectrum of 9-PAH$^+$ (Fig. 7.19b) and the SERS spectrum at the pH = 1.4 (Fig. 7.19c) one can observe that there is only a small shift of the SERS bands to lower wavenumber values by 1–3 cm^{-1} and the bandwidths are hardly affected. These facts support the assumption that a 9-PAH$^+$-chloride ion pair is adsorbed on the silver surface via the chloride ion.

7.3.3 Conclusions

SERS spectra of the 9-PA and 9-PAH$^+$ cation on silver colloidal suspension have been recorded and discussed in comparison with the corresponding Raman spectra. The analysis of the SERS spectra of neutral 9-PA reveals the adsorption of 9-PA on the silver surface via the lone pair electrons of the nitrogen atom. At acidic pH values, the 9-PAH$^+$ cation pairs with the chloride ion and this ion pair is bound to the silver surface via the chloride ion.

References

Albert A (1966) The acridines. Edward Arnold, London

Allen G, Bernstein HJ (1955) Internal rotation: VIII. The infrared and Raman spectra of furfural. Can J Chem 33:1055–1061

Bolboaca M, Iliescu T, Kiefer W (2002) Fourier transform Raman and surface-enhanced Raman spectroscopy of some quinoline derivatives. J Raman Spectrosc 33:207–212

Bunding KA, Bell ML (1982) Surface-enhanced Raman spectroscopy of pyridine derivatives: Effects of adsorption on electronic structure. Surf Sci 118:329–344

Campion A, Kambhampati P (1998) Surface-enhanced Raman scattering. Chem Soc Rev 27:241–250

Chang RK, Furtak T E (1982) Surface-Enhanced Raman Scattering. Plenum Press, New York

Creighton JA (1983) Surface Raman electromagnetic enhancement factors for molecules at the surface of small isolated metal spheres: The determination of adsorbate orientation from SERS relative intensities. Surf Sci 124:209–219

Creighton JA (1986) The resonance Raman contribution to SERS: Pyridine on copper or silver in aqueous media. Surf Sci 173:665–672

Creighton JA, Blatchford CG, Albrecht MG (1979) Plasma resonance enhancement of Raman scattering by pyridine adsorbed on silver or gold sol particles of size comparable to the excitation wavelength. J Chem Soc Faraday Trans 2: Molec&Chem Phys 75:790–798

Dahlqvist KI, Forsen S (1965) The barrier to internal rotation in 2-furanaldehyde. J Phys Chem 69:4062–4071

Demeunynck M, Charmantray F, Martelli A (2001) Interest of acridine derivatives in the anti-cancer chemotherapy. Current Pharma Design 7:1703–1724

Denny WA (1990) Acridine-based antitumour agents. In: Wilman DEV (ed) The chemistry of antitumour agents, Blackie & Co., London

Denny WA (2002) Acridine derivatives as chemotherapeutic agents. Curr Med Chem 9:1655–1665

Dollish FR, Fateley WG, Bentley FF (1973) Characteristic Raman frequencies of organic compounds. John Wiley & Sons, New York

Dryhurst CG (1977) Electrochemistry of biological molecules. Academic Press, New York

Dumaitre BA, Dodic N, Daugan ACM, Pianetti PMC (1997) Certain isoquinoline derivatives having anti-tumor properties. Patent number: 5663179 Laboratoires Glaxo SA

Foreman MI, Taylor M, Clark C, Devitt H, Hanlon G, Kelly I, Lukowiecki G (1985) Pharmacology and treatment isoquinoline is a possible anti-psoriatic agent in coal tar. British Journal of Dermatology 112:323–328

Gao P, Weaver MJ (1985) Surface-Enhanced Raman spectroscopy as a probe of adsorbate-surface bonding: Benzene and monosubstituted benzenes adsorbed at gold electrodes. J Phys Chem 89:5040–5046

Gao X, Davies JP, Weaver MJ (1990) A test of surface selection rules for surface-enhanced Raman scattering: the orientation of adsorbed benzene and monosubstituted benzenes on gold. J Phys Chem 94:6858–6865

Hallmark VM, Campion A (1986) Selection rules for surface Raman spectroscopy: experimental results. J Chem Phys 84:2933–2941

Iliescu T, Marian I, Misca R, Smarandache V (1994) Surface-enhanced Raman spectroscopy of 9-phenylacridine on silver sol. Analyst 119:567–570

Iliescu T, Bolboaca M, Pacurariu R, Maniu D, Kiefer W (2003) Raman spectroscopy, surface-enhanced Raman spectroscopy and density functional theory studies of 2-formylfuran. J Raman Spectrosc 34:705–710

Iliescu T, Vlassa M, Caragiu M, Marian I, Astilean S (1995) Raman study of 9-methylacridine adsorbed on silver sol. Vib Spectrosc 8:451–456

Iliescu T, Irimie FD, Bolboaca M, Paisz Cs, Kiefer W (2002a) Vibrational spectroscopic investigations of 5-(4-fluor-phenyl)-furan-2-carbaldehyde. Vib Spectrosc 29:235–239

Iliescu T, Irimie FD, Bolboaca M, Paisz Cs, Kiefer W (2002b) Surface enhanced Raman spectroscopy of 5-(4-fluor-phenyl)-furan-2 carbaldehide adsorbed on silver colloid. Vib Spectrosc 29:251–255

Iliescu T, Irimie FD, Bolboaca M, Paisz Cs, Kiefer W (2001) Raman spectra of 5 substituted-furan-2-carbaldehyde adsorbed on silver sol, Book of Abstracts of the 1st International Conference on Advanced Vibrational spectroscopy (ICAVS), Turku, P14171

Iliescu T, Irimie FD, Baia M, Paizs Cs, Bratu I, Kiefer W (2003–2004) IR absorption, FT-Raman and SERS investigations together with DFT calculations on a furan-2-carbaldehyde derivative. Asian Chem Lett 6–7:213–218

Iliescu T, Bolboaca M, Cinta-Pinzaru S, Pacurariu R, Maniu D, Ristoiu M, Kiefer W (2002c) Raman, surface-enhanced Raman spectroscopy and density functional theory studies of 2-formylfuran, Proceedings of the XVIIIth International Conference on Raman Spectroscopy (ICORS), Budapest, 311–312

Ishikawa K, Hayama T, Nishikibe M, Yano M (1994) Isoquinoline derivatives. Patent number 5362736 Banyu Pharmaceutical Co. Ltd

Johansson G, Sundquist S, Nordvall G, Nilson BM, Brisander M, Nilvebrant L, Hacksell U (1997) Antimuscarinic 3-(2-furanyl)quinuclidin-2-ene derivatives: Synthesis and structure-activity relationships. J Med Chem 40:3804–3819

Karabatsos GJ, Vane FM (1963) Structural studies by nuclear magnetic resonance. VI. The stereospecificity of coupling between protons separated by five bonds and conformations of some aromatic aldehydes. J Am Chem Soc 85:3886–3888

Katritzky AR (1963) Physical methods in heterocyclic chemistry vol. II. Academic Press, New York

Kato TA, Hakura A, Mizutani T, Saeki KI (2000) Anti-mutagenic structural modification by fluorine-substitution in highly mutagenic 4-methylquinoline derivatives. Mutation Research/Genetic Toxicology and Environmental Mutagenesis. 465:173–182

LaVoie EJ, Adams EA, Shigematsu A, Hoffman D (1983) On the metabolism of quinoline and isoquinoline: possible molecular basis for differences in biological activities. Carcinogenesis 4:1169–1173

Leite JPV, Oliveira AB, Lombardi JA, Filho JDS, Chiari E (2006) Trypanocidal activity of triterpenes from *Arrabidaea triplinervia* and derivatives. Biol Pharm Bull 29:2307–2309

Levi G, Pantigny J, Marsault JP, Christensen D H, Nielsen OF, Aubard J (1992) Surface-enhanced Raman spectroscopy of ellipticines adsorbed onto silver colloids. J Phys Chem 96:926–931

Liu LF (1989) DNA topoisomerase poisons as antitumor drugs. Annu Rev Biochem 58:351–375

Lombardi JR, Birke RL, Lu T, Xu J (1986) Charge-transfer theory of surface enhanced Raman spectroscopy: Herzberg–Teller contributions. J Chem Phys 84:4174–4180

Miller FA, Fateley WG, Witkowski RE (1967) Torsional frequencies in the far infrared—V. Torsions around the C---C single bond in some benzaldehydes, furfural, and related compounds. Spectrochim Acta Part A: Molec Spectrosc 23:891–908

Moskovits M (1985) Surface-enhanced spectroscopy. Rev Mod Phys 57:783–826

Moskovits M, DiLella DP (1980) Surface-enhanced Raman spectroscopy of benzene and benzene-d_6 adsorbed on silver. J Chem Phys 73:6068–6075

Moskovits M, Suh JS (1984) Surface selection rules for surface enhanced Raman spectroscopy: calculations and applications to surface-enhanced Raman spectrum of phthalazine on silver. J Phys Chem 88:5526–5530

Moskovits M, Suh JS (1988) Surface geometry change in 2-naphthoic acid adsorbed on silver. J Phys Chem 92:6327–6329

Mukherjee K, Bhattacharjee D, Misra TN (1997) Surface-enhanced Raman spectroscopic study of isomeric formylthiophenes in silver colloid. J Colloid Interface Sci 193:286–290

Oh ST, Kim K, Kim MS (1991) Adsorption and surface reaction of acridine in silver sol: surface-enhanced Raman spectroscopic study. J Phys Chem 95:8844–8849

Park SM, Kim K, Kim MS (1994) Raman spectroscopy of isonicotinic acid adsorbed onto silver sol surface. J Molec Struct 328:169–178

Ribero-Claro PJA, Batista de Carvalho LAE, Amado AM (1997) Evidence of dimerization through C – H···O interactions in liquid 4-methoxybenzaldehyde from Raman spectra and ab initio calculations. J Raman Spectrosc 28:867–872

Rogers DJ, Luck SD, Irish DE, Guzonas DA, Atkinson GF (1984) Surface enhanced Raman spectroscopy of pyridine, pyridinium ions and chloride ions adsorbed on the silver electrode. J Electroanal Chem 167:237–249

Sanchez-Cortes S, García-Ramos JV (1990) Surface-enhanced Raman spectroscopy of 1,5-dimethylcytosine on silver and copper sols. J Raman Spectrosc 21:679–682

Sun SC, Bernard I, Birke RL, Lombardi JR (1985) The effect of pH, chloride ion and background electrolyte concentration on the SERS of acidified pyridine solutions. J Electroanal Chem 196:359–374

Tanaka A, Terasawa T, Hagihara H, Sakuma Y, Ishibe N, Sawada M, Takasugi H, Tanaka H (1998) Inhibitors of acyl-CoA:cholesterol O-acyltransferase. 2. Identification and structure-activity relationships of a novel series of N-alkyl-N-(heteroaryl-substituted benzyl)-N'-arylureas, J Med Chem 41:2390–2410

Vo-Dinh T (1988) Surface-enhanced Raman spectroscopy using metallic nanostructures. Trends Analyt Chem 17:557–582

Wainwright M (2001) Acridine-A neglected antimicrobial chromophore. J Antimicrobial Chemotherapy 47:1–13

Wait SC, McNerney JC (1970) Vibrational spectra and assignments for quinoline and isoquinoline. J Molec Spectrosc 34:56–77

Werbel LM, Steck EA (1987) Anti-leishmanial lepidine derivatives. Patent number ADD012795

8 New Developments in SERS-Active Substrates

As already mentioned, in the last ten years the interest in SERS was completely reconsidered as a consequence of the progress attained in nanoscience and nanotechnology. In recent times, there is an astonishing research interest concerning how to control, manipulate, and amplify light on the nanometer length scale using the surface plasmons properties. Thus, a great variety of surface-confined nanostructures were already produced by numerous fabrication methods, including colloid immobilization (Li et al. 2004, Wang and Gu 2005, Orendorff et al. 2005, Hu et al. 2007, Li et al. 2007, Wei et al. 2007, Zhou et al. 2007, Zhou et al. 2006), electron-beam lithography (Gunnarsson et al. 2001, Grand et al. 2005, Grand et al. 2003, Felidj et al. 2004, Felidj et al. 2002a) and nanosphere lithography (Haynes et al. 2005, Haynes and Van Duyne 2002, Haynes and Van Duyne 2003, Schmidt et al. 2004, McFarland et al. 2005). However, researchers continue to develop novel SERS substrates to prolong substrate lifetime, to provide stable and optimized enhancement factors, and to permit SERS studies in diverse environments.

All described SERS investigations were performed by employing a silver or gold colloidal suspension as a SERS-active substrate. Having in mind that the molecular species have a specific adsorption behavior on different metallic surfaces (silver or gold) the next step would be the analysis of the adsorption behavior of such molecules on metallic nanostructured substrates with different morphologies. Having in view the relationship between the substrate plasmonic response, the particularities of the investigated compounds, and the available excitation laser line, further insights into the adsorption behavior can be obtained by using various types of SERS-active substrates (ordered (reprinted with permission from J. Phys. Chem. B 2006, 120, 23982–23936, copyright 2006 American Chemical Society and with permission from Appl. Phys. Lett., 2006, 88, 1431211-4, copyright 2006 American Institute of Physics) or disordered (reprinted from Chem. Phys. Lett., 422, Baia M, Toderas F, Baia L, Popp J, Astilean S, Probing the enhancement mechanisms of SERS with p-aminothiophenol molecules adsorbed on self-assembled gold colloidal nanoparticles, 127–132, copyright 2006, with permission from Elsevier and from Nanotechnology, Toderas F, Baia M,

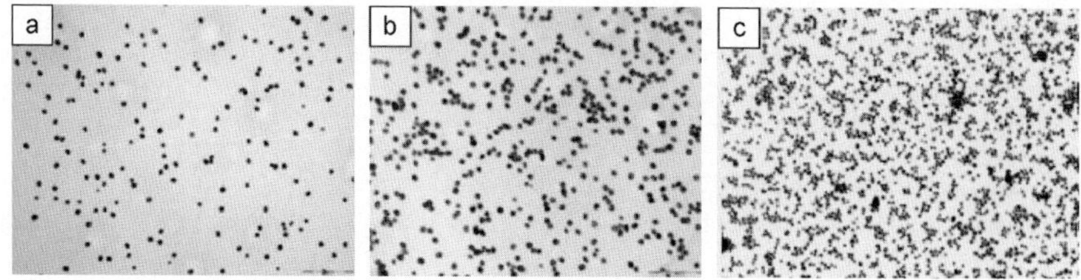

Fig. 8.1 TEM pictures of the gold nanoparticle assemblies formed upon functionalized glass slides by immersing the substrate in colloidal suspension for different time intervals: (**a**) 5, (**b**) 60, (**c**) 120 min (Nanotechnology, Toderas F, Baia M, Baia L, Astilean A, Controlling gold nanoparticle assemblies for efficient surface enhanced Raman scattering (SERS) and localized surface plasmon resonance (LSPR) sensors, 2007, 18, 255702. Reproduced by permission of IOP Publishing Ltd.)

Baia L, Astilean A, Controlling gold nanoparticle assemblies for efficient surface enhanced Raman scattering (SERS) and localized surface plasmon resonance (LSPR) sensors, 2007, 18, 255702, with permission of IOP Publishing Ltd.) metallic nanostructures). In this respect, some substrates have already been prepared in our laboratory by the self-assembling of colloidal nanoparticles on functionalized substrates (Baia et al. 2006a, Toderas et al. 2004a, Toderas et al. 2004b, Baia et al. 2005a, Baia et al. 2006b, Toderas et al. 2006a, Toderas et al. 2006b, Astilean et al. 2006a, Toderas et al. 2007) and nanosphere lithography (Bolboaca et al. 2003a, Astilean et al. 2004a, Bolboaca et al. 2004b, Astilean et al. 2003, Astilean et al. 2004b, Baia et al. 2005b, Astilean et al. 2005, Baia et al. 2006c, Baia et al. 2006d, Astilean et al. 2006b, Baia et al. 2006e) and their SERS efficiency was tested and optimized by using probe molecules.

Gold colloidal nanoparticles were immobilized on functionalized glass substrates and their efficiency as SERS-active substrates was tested for different nanoparticle density (Toderas et al. 2007) and excitation lines (Baia et al. 2006a). The TEM pictures of the gold nanoparticle assemblies formed upon functionalized glass slides by immersing the substrate in colloidal suspension for different time intervals are depicted in Fig. 8.1.

According to the image analysis produced by using an image processing toolkit, the mean value of the nanoparticle diameter was 18 ± 2 nm and the average number of particles captured on glass surface was 147, 585 and 1109 nanoparticles/μm^2 for 5, 60, and 120 min immersion time, respectively. Adhesion of gold particles to the substrate is due to interfacial electrostatic interactions between the positively charged, protonated terminal amino groups of aminopropyltriethoxysilane from the glass substrate and the citrate anions attached to the gold nanoparticles. As expected, for a short time of exposition, colloids tend to adsorb as isolated particles and, rarely, as dimers or trimers. However, with increasing the immersion time the glass surface coverage increases as well, the particles exhibit a certain degree of compactness and aggregation. In order to account for the surface coverage and the degree of aggregation Table 8.1 gives the distribution of different assemblies (single, dimer, trimer, and cluster) formed on the substrates.

The absorption spectra of samples presented in Fig. 8.1 are depicted in Fig. 8.2.

Table 8.1 The number of nanoparticles and assemblies captured on the substrate after (**a**) 5, (**b**) 60 and (**c**) 120 min immersion time (Nanotechnology, Toderas F, Baia M, Baia L, Astilean A, Controlling gold nanoparticle assemblies for efficient surface enhanced Raman scattering (SERS) and localized surface plasmon resonance (LSPR) sensors, 2007, 18, 255702. Reproduced by permission of IOP Publishing Ltd.)

Nanoparticles/assembly	Sample		
	(a)	(b)	(c)
1	154	218	187
2	9	50	60
3	5	22	48
4	0	5	25
5	0	10	13
6–10	0	7	33
>10	0	3	68

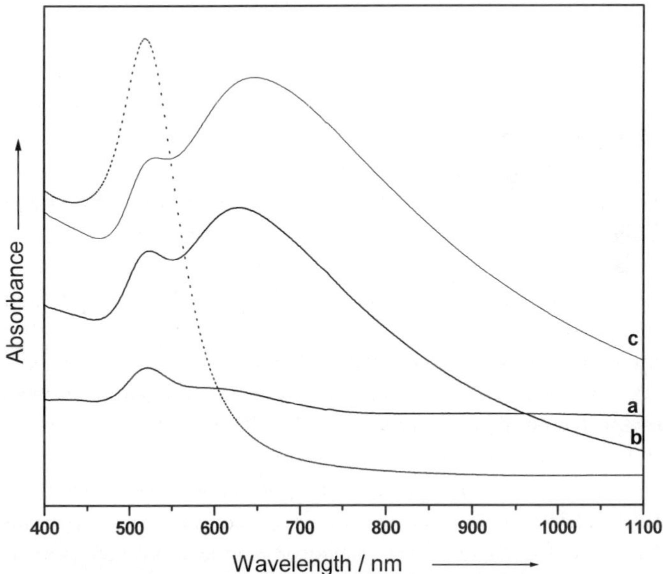

Fig. 8.2 Modification of UV-vis absorption spectra of the gold nanoparticle assemblies formed upon functionalized glass substrates as a result of different time intervals of immersion in colloidal suspension: (*a*) 5 min, (*b*) 60 min and (*c*) 120 min. A reference spectrum of gold nanoparticles in aqueous solution is included (dotted line) (Nanotechnology, Toderas F, Baia M, Baia L, Astilean A, Controlling gold nanoparticle assemblies for efficient surface enhanced Raman scattering (SERS) and localized surface plasmon resonance (LSPR) sensors, 2007, 18, 255702. Reproduced by permission of IOP Publishing Ltd.)

When compared to the spectrum of an aqueous solution of colloidal gold, the UV-vis absorption spectra of gold nanoparticles immobilized on the glass substrate display two plasmon bands. The short wavelength band was assigned to the intrinsic plasmon resonance of isolated particles and the long wavelength band to a collective surface plasmon oscillation. The maximum of the short wavelength plasmon band is shifted from 520 nm for an isolated particle in an aqueous solution to 528 nm, as expected due to the greater dielectric constant of the glass substrate than that of water. The long wavelength band at around 635 nm is developing as a function of the immersion time and reflects the local density of particles and the degree of their assembling on the substrate, as revealed by TEM pictures. When gold nanoparticles are located near to each other, the individual plasmon resonances couple to each other via the near field, shifting to red the local collective plasmon resonance. The electromagnetic coupling between particles in disordered and organized metallic arrays has been widely investigated both experimentally and theoretically and the mechanism is well established (Baia et al. 2006a, Atay et al. 2004, Tamaru et al. 2002, Aizpurua et al. 2005). Note that the UV-vis absorption is particularly strong at the excitation wavelength (633 nm) which recommends the samples of high particle density to be used as an SERS substrate.

The Raman spectrum of solid p-aminothiophenol (p-ATP) that was employed as a probe molecule and the representative SERS spectra of adsorbed p-ATP molecules are shown in Fig. 8.3. It is worth mentioning that many similar SERS spectra were collected from different points on each sample and their Raman intensity and band positions do not vary significantly when the laser spot is moved laterally.

By calculating the average enhancement factors (EFs) of the Raman signal they were found to be in the range of 10^5–10^7, namely 2×10^5 for the lowest gold nanoparticles density, 1.7×10^6 for the substrate maintained 60 min into colloidal suspension and 5×10^7 for the highest nanoparticles density (120 min immersion time) (Toderas et al. 2007). It is obviously from these values that the enhancement gain by factors of 8.5 and 250 (relative to the lowest coverage substrate) does not go exactly with the number of particles, but more likely with the number of assemblies and interstitial sites in assemblies (Table 8.1).

It is well known that the enhancement of the local field can vary by several orders of magnitude from isolated to dense packing particles. In addition, SERS is a highly heterogeneous process due to local enhancements at "hot spots" and many experimental results in the literature, particularly on SERS from single molecules on aggregated nanoparticles, are interpreted on the basis of the "hot spot" mechanism (Tian et al. 2002, Felidj et al. 2004). Along with large fractal aggregates cited in the literature (Moskovits 2005, Campion and Kambhampati 1998), in which the "hot spots" arise from the symmetry breaking, other classes of particle assemblies as dimers or small aggregates are capable of producing "hot sites." On the other hand, organized arrays of metallic nanoparticles fabricated through electron beam lithography can also sustain huge electromagnetic field confined around the particles, leading to very effective SERS substrates (Grand et al. 2003, Grand et al. 2005, Billot et al. 2006).

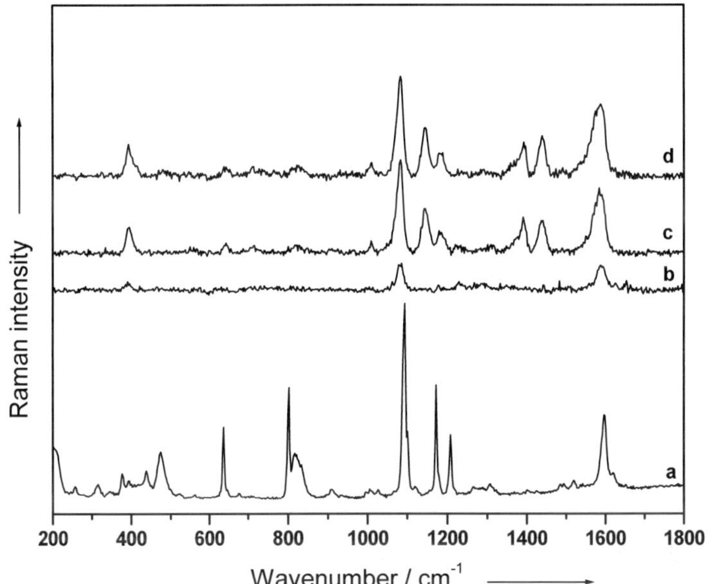

Fig. 8.3 The normal Raman spectrum of solid p-ATP (*a*) together with the SERS spectra of saturated monolayer of p-ATP molecules adsorbed on gold nanoparticles self-assembled on the glass substrate (*b*) after 5 min (*c*) 60 min and (*d*) 120 min immersion time (Nanotechnology, Toderas F, Baia M, Baia L, Astilean A, Controlling gold nanoparticle assemblies for efficient surface enhanced Raman scattering (SERS) and localized surface plasmon resonance (LSPR) sensors, 2007, 18, 255702. Reproduced by permission of IOP Publishing Ltd.)

Encouraged by the plasmonic response of the substrate maintained into the gold colloidal suspension for 120 min with adsorbed p-ATP molecules (see Fig. 8.4) our further interest was to assess the enhancement potential of the substrate by using different excitation laser lines. The selected laser lines excite both the surface plasmon resonances of the isolated gold nanoparticles (532 nm) as well as those of the clusters formed on the glass substrate (633 and 830 nm). The 1064 nm laser line, which was also used for recording the SERS spectra, is located far away from the maximum of the plasmonic resonance band of the gold particles.

Getting SERS with excitation in near-infrared region would considerably increase the applicability areas of this substrate, making it suitable for the investigation of various molecules of pharmacological, biological, and medical interest.

The SERS spectra of p-ATP molecules adsorbed on this substrate and recorded with four different laser lines from visible and near-infrared spectral regions are illustrated in Fig. 8.5 (Baia et al. 2006a, Baia et al. 2006b).

The different enhancement patterns evidenced in the SERS spectra recorded at these excitation lines were explained from the perspective of the different contributions of the mechanisms responsible for the total SERS enhancement. The interplay of the enhancement mechanisms' contribution to the overall enhancement was further confirmed by quantitatively evaluating the areas ratio of the

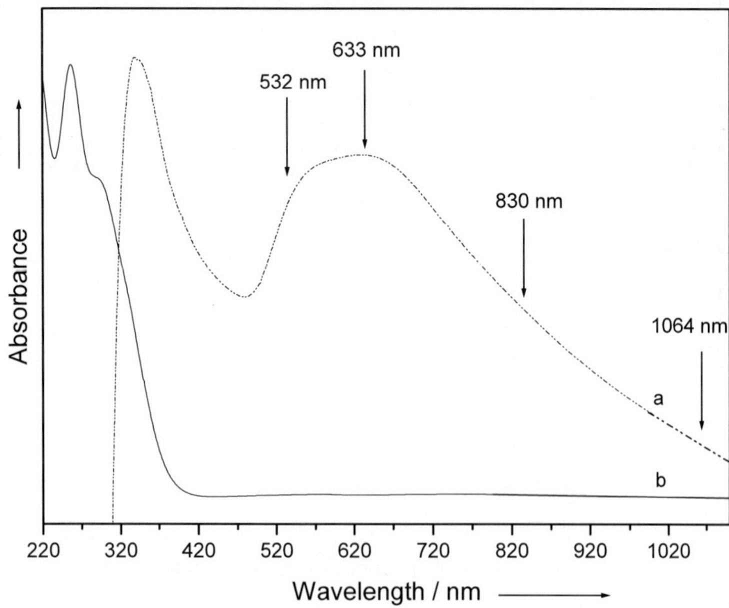

Fig. 8.4 UV-vis absorption spectra of the SERS substrate with adsorbed p-ATP molecules (*a*). The absorption spectrum of p-ATP solution in MeOH is also presented (*b*). Reprinted from Chem. Phys. Lett., 422, Baia M, Toderas F, Baia L, Popp J, Astilean S, Probing the enhancement mechanisms of SERS with p-aminothiophenol molecules adsorbed on self-assembled gold colloidal nanoparticles, 127–132, copyright 2006, with permission from Elsevier.

bands enhanced by electromagnetic and charge-transfer mechanisms located around 1080 and 1433 cm^{-1}, respectively, from the spectra recorded with 532, 633, and 830 nm. Thus, on passing from visible to near-infrared excitation the ratio values considerably increased and were found to be 1.55, 2.32, and 13.5, respectively.

Thus, the as prepared self-assembled gold nanoparticles turn out to be not only a model for gaining further understanding in SERS mechanisms but also a promising substrate for other SERS-based measurements in near-infrared. The possibility of getting good SERS spectra with near infrared excitations opens up promising perspectives for the investigations of biological samples.

Others metallic nanostructures were obtained in order to manipulate their potential as SERS-active substrates and consist of corrugated gold films deposited on top of highly ordered polystyrene nanospheres arrays (Baia et al. 2005b, Astilean et al. 2005, Baia et al. 2006c, Baia et al. 2006d, Astilean et al. 2006, Baia et al. 2006e). Figure 8.6 shows a $10 \times 10\,\mu m^2$ image of the 60 nm nanostructured gold film, together with its section analysis. Beside this film other two nanostructured gold films of 30 and 15 nm were deposited on the polystyrene spheres.

It is remarkable to note the existence of large domains of 2D colloidal crystals. The majority of the self-assembled nanospheres have a hexagonally close-packed

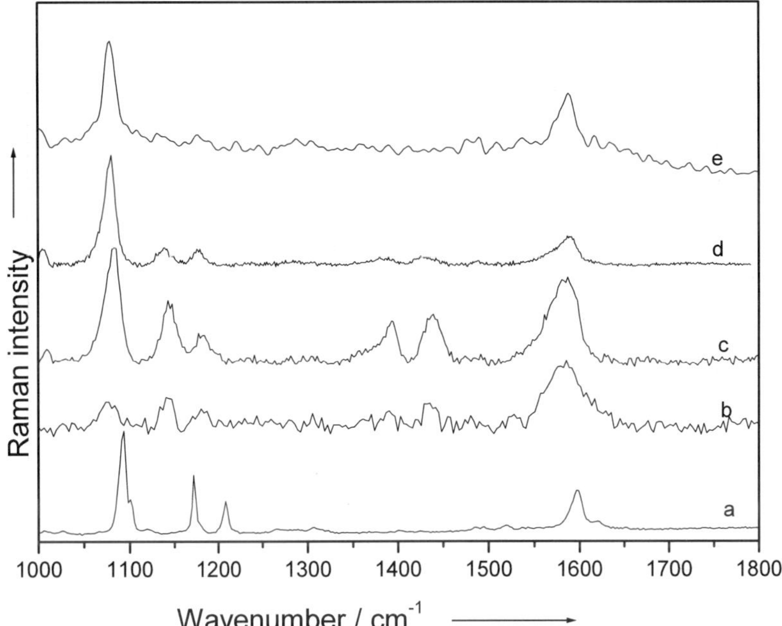

Fig. 8.5 The normal Raman spectrum of solid p-ATP recorded with 633 nm laser line (*a*) and the SERS spectra of p-ATP adsorbed on self-assembled gold colloidal nanoparticles recorded with different laser lines: (*b*) 532 nm, (*c*) 633 nm, (*d*) 830 nm, (*e*) 1064 nm. Reprinted from Chem. Phys. Lett., 422, Baia M, Toderas F, Baia L, Popp J, Astilean S, Probing the enhancement mechanisms of SERS with p-aminothiophenol molecules adsorbed on self-assembled gold colloidal nanoparticles, 127–132, copyright 2006, with permission from Elsevier

symmetry. However, a few features like linear and local dislocations, which usually appear on such nanosphere masks, can be seen. Nevertheless, these characteristics should not considerably affect the reproducibility of the SERS spectra.

One should also emphasize that the surface morphology of the fabricated metallic films is quite complex, due to the superimposition of two gratings consisting of half-shells and truncated tetrahedra that are formed in the spaces between the spheres. The latter are highlighted in the inset from the upper left corner of Fig. 8.6.

To probe the plasmonic response of the investigated substrates, reflectance measurements have been performed and those recorded at 50° off normal incidence are presented in Fig. 8.7. As one can see, the reflectance spectra show strong reflectivity around 540 and 800 nm separated by pronounced reflectivity dips around 350, 440, 640, and 840 nm. The existence of two spatially separated metallic gratings, one arising from the obvious half-shells and the other one from less obvious truncated tetrahedra, which are geometrically different and give their own optical responses, as well as the scattering contribution of the polystyrene

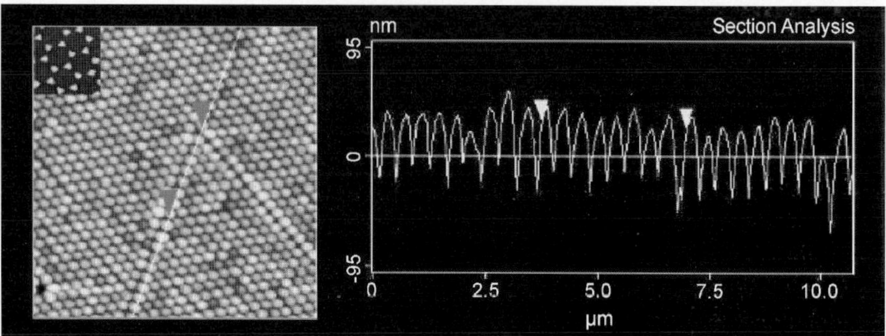

Fig. 8.6 AFM image of a $10 \times 10\,\mu m^2$ region of the 60 nm gold nanostructured substrate together with its section analysis. The inset shows the truncated tetrahedra formed in the spaces between spheres (400 nm diameter) at an expanded scale. Reprinted with permission from J. Phys. Chem. B 2006, 120, 23982–23936, copyright 2006 American Chemical Society

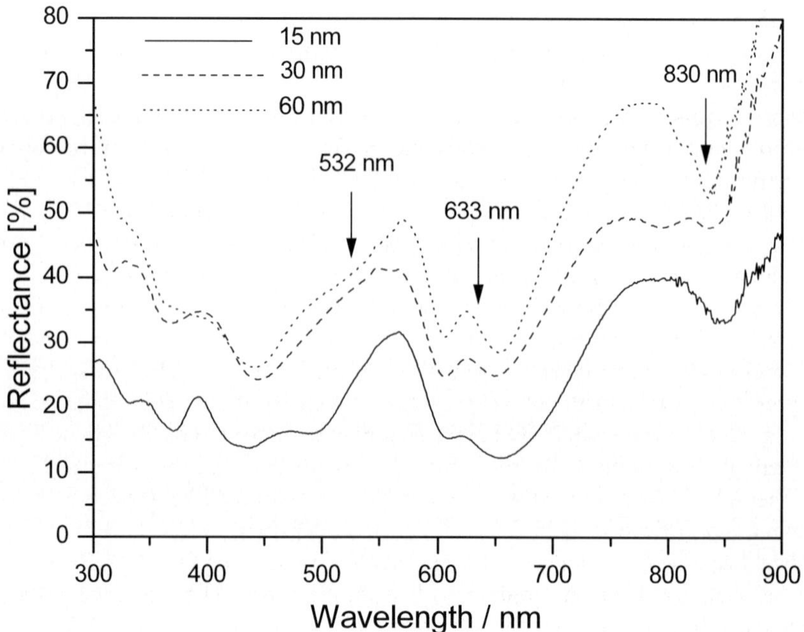

Fig. 8.7 Reflectance spectra of gold nanostructured substrates of different thicknesses as depicted recorded at 50° off normal incidence. The arrows indicate the employed excitation laser lines. Reprinted with permission from J. Phys. Chem. B 2006, 120, 23982–23936, copyright 2006 American Chemical Society

spheres substrate, make difficult the accurate analysis of the overall optical response of these nanostructured films.

Similar reflectivity minima were observed around 525 and 650 nm in the spectra, recorded on arrays of gold oblate spheroidal particles deposited onto a 20 nm thick gold film, and were assigned to the mixed modes proceeding from the simultaneous excitation of localized and propagating surface plasmon modes, and from the excitation of an ensemble of particles strongly coupled by propagating surface plasmon wave, respectively (Felidj et al. 2002b).

Taking into account previously reported results, (Baia et al. 2005b) where optical properties of 54 nm gold films deposited on top of highly ordered polystyrene nanosphere (220 nm) arrays were investigated in relation to those of the flat metallic film, in this case it was assumed that the optical characteristics at wavelength values lower than 700 nm are due to the simultaneous excitation of both localized plasmons, supported by quasi-isolated gold nanoprisms, and the propagating ones, from both metal dielectric interfaces. The change of the reflectivity dip positions with increasing film thickness can be explained by assuming that the surface waves on two metal-dielectric interfaces are resonant at different wavelength values and that additionally they could interfere constructively or destructively.

This behavior can consequently imply that certain thicknesses are efficient either in transmission or in reflectance at a specific wavelength. This would provide further insights concerning the coupling between the localized and propagating plasmons, and would certainly contribute to a deep understanding of the optical properties of the fabricated gold nanostructured films.

In the near-infrared spectral domain of the reflectance spectra (Fig. 8.7) one observes a slight shift to lower wavelengths of the reflectance dip in the 830 nm wavelength range with the progressive increase of the film thickness. Having in view that strongly localized electromagnetic fields exist both at the top of the spheres and between them, we assume that the observed dip is most probably due to the electromagnetic coupling between the interconnected gold half-shells. The progressive increase of the reflectance dip as the gold film becomes thicker could lead to the electromagnetic field confinement either between calottes or at the top of them. If this is the case, the gold nanostructured substrate with the highest thickness should consequently provide the greatest SERS enhancement under near-infrared excitation.

The above analysis suggests that by using a given laser line, the SERS signal can be maximized for an optimum gold film thickness. Therefore, the enhancement capabilities of the prepared substrates with laser lines covering a wide wavelength domain from visible to near-infrared were effectively tested.

The recorded SERS spectra of the adsorbed p-ATP molecules are presented in Fig. 8.8 together with their corresponding Raman spectra. As revealed by Fig. 8.8, with increasing the thickness of the deposited gold films (15, 30, and 60 nm) different average EF were obtained depending on the employed laser lines.

Thus, one can see (Baia et al. 2006e) the existence of a more than 220 times gain of the average enhancement of the Raman signal of the p-ATP molecules

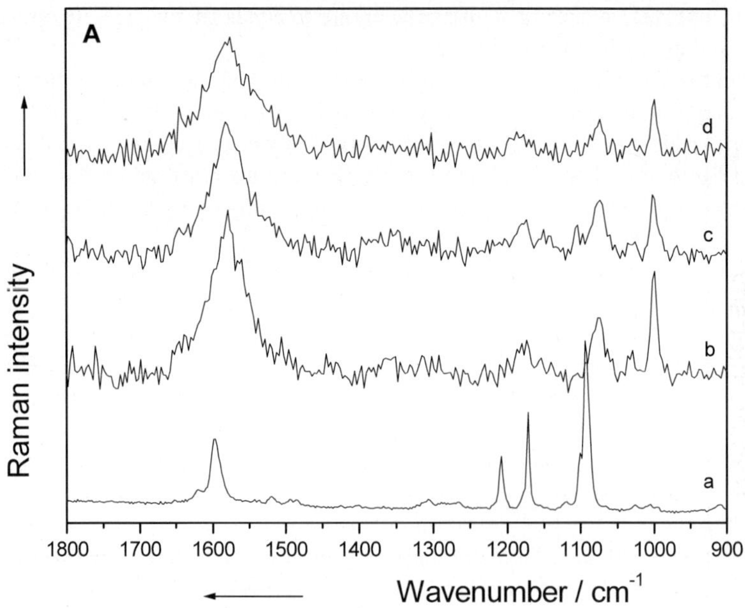

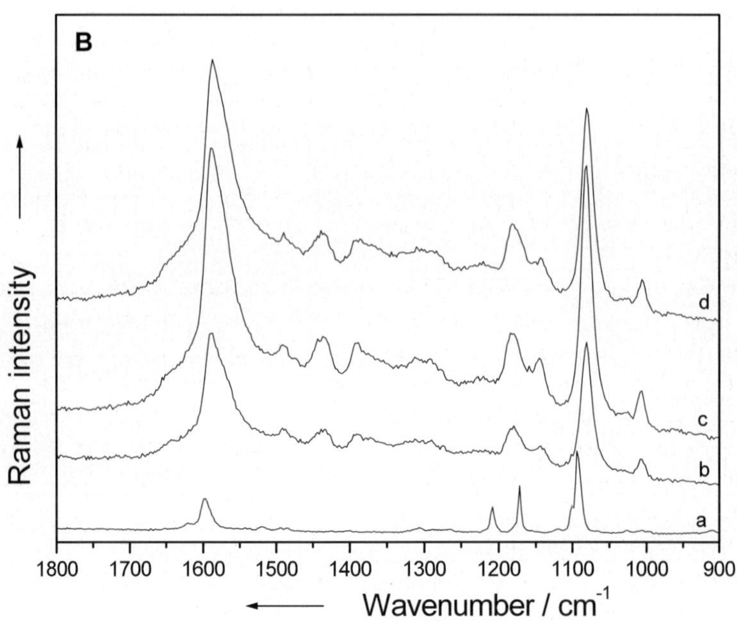

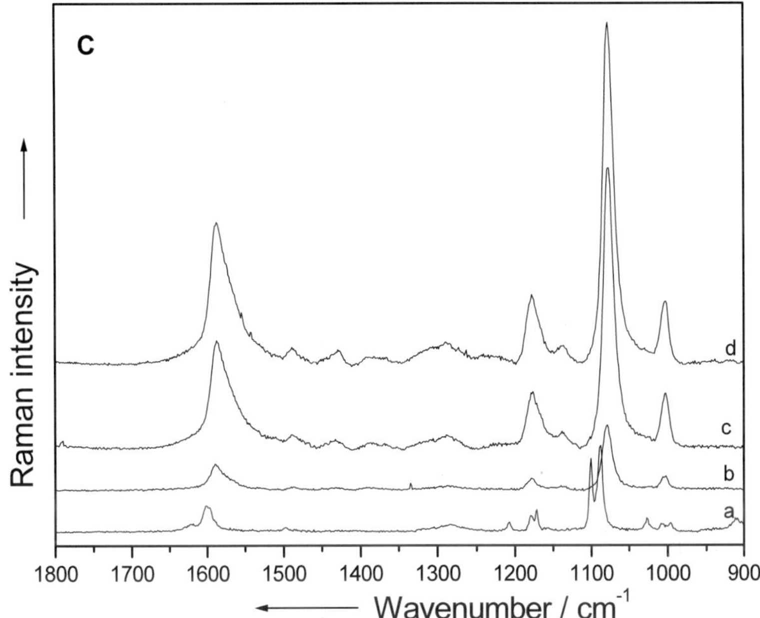

Fig. 8.8 The normal Raman spectra of solid p-ATP (*a*) and the SERS spectra of p-ATP adsorbed on gold nanostructured films with thickness of 15 nm (*a*), 30 nm (*b*) and 60 nm (*c*) recorded with 532 (**A**), 633 (**B**) and 830 (**C**) nm excitation laser lines. The SERS spectra recorded with a 532 nm line were baseline corrected. Reprinted with permission from J. Phys. Chem. B 2006, 120, 23982–23936, copyright 2006 American Chemical Society

adsorbed on the 60 nm nanostructured film when an 830 nm line was used for excitation (1.23×10^6), in comparison with that exhibited under 532 nm excitation (5.4×10^3). Moreover, for the 830 nm laser line one also notes the significant increase of the average EF for the 30 nm gold nanostructured film. By comparing the average EF values obtained from different substrates excited with the same laser line, the existence of a wavelength-dependent enhancement can be observed (see Fig. 8.9).

Thus, for the excitation with 532 nm the most enhanced SERS signal was obtained from the sample with 15 nm thickness, while for excitation with 830 nm the most enhanced SERS signal was achieved from the sample with 60 nm thickness. When the 633 nm line was used for excitation, the most enhanced SERS signal was recorded from the sample of 30 nm thickness, due probably to a stronger coupling between surface plasmons at this thickness. The major differences of the average EF values were obtained for the gold nanostructured films when an 830 nm excitation line was employed. The large tunability of surface plasmon excitation combined with the advantage of relatively high exhibited average EF values recommends these substrates as outstanding candidates for upcoming investigations of various pharmaceutically, biologically, and medically relevant molecules.

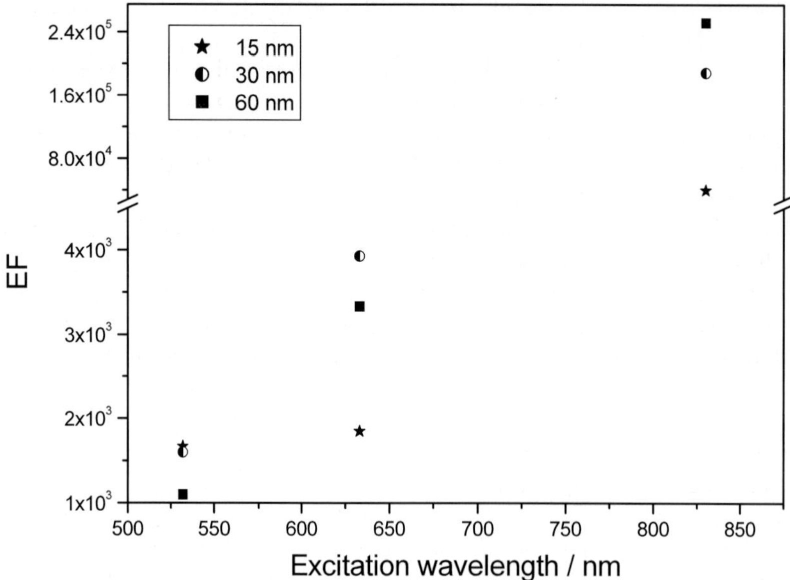

Fig. 8.9 The calculated average enhancement factor as a function of the excitation laser line. Reprinted with permission from J. Phys. Chem. B 2006, 120, 23982–23936, copyright 2006 American Chemical Society

The SERS efficiency of truncated tetrahedral silver nanoparticles arrays with different vertical dimensions prepared by nanosphere lithography was also assessed. Triangular silver nanoparticle arrays present a significant advantage over many of the traditional SERS substrates because they exhibit narrow size distributions and their localized surface plasmon resonance (LSPR) can be tuned throughout the visible and near-infrared wavelengths by systematically varying the nanoparticles dimensions (Haynes and Van Duyne 2003, McFarland et al. 2005).

Recently, it was shown (Schmidt et al. 2004) that the SERS of rhodamine 6G (R6G) on 90 and 200 nm triangularly shaped silver nanocluster arrays prepared by nanosphere lithography exhibit an order of magnitude enhancement gain relative to the amorphous silver film with the same thickness (10 nm) due to the large electromagnetic fields located at the nanoclusters edges and corners, the 90 nm nanoclusters providing the strongest enhancement. Substrate SERS efficiency was found to increase rapidly with decreasing interparticle separation, signalling the importance of strong interparticle couplings effects in SERS (Gunnarsson et al. 2001). Theoretical simulations (Zou and Schatz 2005) described that arrays of dimmers composed of truncated tetrahedral silver nanoparticles produce giant enhancements in electromagnetic fields at particles' edges and tips due to the mixing between the particle LSPR and long range photonic interactions. Moreover, it was shown that a variation of the particle size from a 100 nm in-plane width and a 30 nm vertical height to 167 and 50 nm, respectively, leads to an enhancement of the electromagnetic field by a factor of ~6 due to the increase in the long range coupling effect.

To gain further insights into the enhancement capabilities of nanosphere lithography prepared substrates, we evaluated the SERS efficiency of R6G adsorbed on triangular silver nanoclusters arrays with the same in-plane width (90 nm) and different vertical dimensions (50 and 70 nm) by using 532 and 633 nm excitation laser lines.

The silver nanoclusters morphology was explored with AFM in taping mode and their topographical image, together with the vertical profile along the indicated direction, are shown in Fig. 8.10. The silver nanoparticles resemble as expected (Haynes and Van Duyne 2003) a truncated tetrahedron with an in-plane width of 90 nm as determined from geometrical considerations (Hulteen and Van Duyne 1995) and out-of-plane heights of 50 and 70 nm. Besides the well-defined edges of the hexagonally packed triangles (see Fig. 8.10) one can see that they form geometrical ordered arrays, which could favor the occurrence of the coherent electromagnetic couplings between the localized and long range modes. A few other features are observed on the SERS substrate. The continuous metallic wires result from not completely close packed nanospheres that allow the metal to penetrate between them. The dark regions appear as a consequence of the silver nanoclusters lifting off from the substrate during the sonication process. These features represent, however, a few percents of the surface and are recognised as characteristic defects of nanosphere masks (Schmidt et al. 2004, Zhang et al. 2005).

In order to avoid their influence on the overall Raman enhancement, and to ensure the substrates' reproducibility, SERS spectra were recorded from multiple sites on the substrates surface. For each laser line and tested substrate similar SERS spectral characteristics, i. e., enhancement, position, and relative intensity of the bands, were obtained, thus proving the good reproducibility of these substrates.

The SERS spectra of the R6G molecules adsorbed on the as-prepared silver triangular arrays are shown in Fig. 8.11*a*.

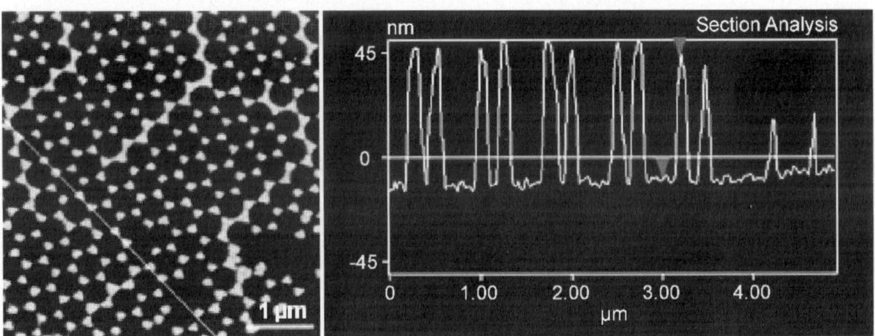

Fig. 8.10 AFM micrograph ($5 \times 5\,\mu m^2$) of the 90 nm clusters obtained from 400 nm polystyrene spheres with an out-of-plane height of 70 nm together with their section analysis. Reprinted with permission from Appl. Phys. Lett., 2006, 88, 1431211-4, copyright 2006 American Institute of Physics

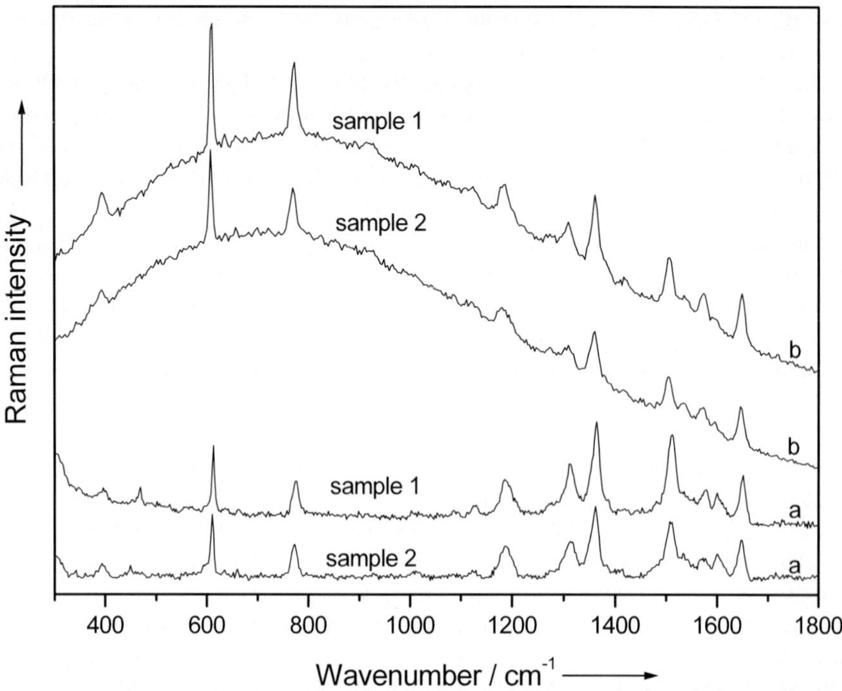

Fig. 8.11 SERS (*a*) and SERRS (*b*) spectra of the R6G molecules adsorbed on ordered silver nanocluster arrays with an in-plane width of 90 nm and out-of-plane heights of: (sample 1) 50 nm, and (sample 2) 70 nm, respectively. Reprinted with permission from Appl. Phys. Lett., 2006, 88, 1431211-4, copyright 2006 American Institute of Physics

Prior to calculating the absolute EF values of both substrates their relative enhancement was determined by multiplying the SERS intensities ratio for a selected band ($1650\,\text{cm}^{-1}$) with the inverse ratio of the nanoparticles exposed areas. Thus, it was seen that the nanoparticles with a 50 nm height provide a stronger enhancement by 1.34 times, compared to that of the nanoparticles with a 70 nm out-of-plane dimension. Since the nanoparticles exposed areas' ratio value is 1.1 one assumes the existence of a supplementary enhancement, which does not originate from the nanoparticles' geometrical area change. Moreover, the following question arises: is the entire nanoparticles exposed area indeed active? Thus, we tried to record the SERS spectrum of the R6G molecules adsorbed on a 50 nm flat silver film, and no SERS signal was obtained. This proves without any doubt that the obtained enhancement is not due to the nanoclusters' surface roughness. Moreover, it becomes obvious that the SERS enhancement is due to the nanoparticles edges and tips, and consequently, the nanoparticles geometrical area is an overestimation of the effective SERS active surface.

Theoretical simulations have shown that exceptionally large electromagnetic field enhancements occur at specific locations (edges and tips) in the structure of dimmers composed of truncated tetrahedral silver nanoparticles as a result of the mixing between the LSPR of each particle and photonic modes of the particle array (Zou and Schatz 2005).

According to these findings, we estimated that huge electromagnetic fields are located on approximately 7% of the bottom triangle area. Extrapolating this percentage to the entire nanoparticles' exposed area we evaluated the effective SERS active area of the silver nanoclusters and further calculated the absolute EF values. Thus, it was found that both substrates exhibit an EF of an order of 10^7. The nanoparticles with 50 nm height provide an enhancement of 2.5×10^7, while the EF value of the silver nanoclusters with 70 nm height was of 1.9×10^7. Having in mind that the electromagnetic couplings between the particles LSPR and long range photonic interactions in a one-dimensional array are solely accountable for the SERS enhancement of the R6G molecules adsorbed on the as prepared substrates it was concluded that the EF increasing, with decreasing the nanoparticles' height, clearly proves the electromagnetic couplings sensitivity to the nanoparticles' vertical dimension (Baia et al. 2006c).

In order to test the overall validity of the above findings we calculated the substrates' relative enhancement by using the 613 and 1364 cm^{-1} band intensities. Similarly, the calculations revealed that nanoparticles with smaller out-of-plane dimensions exhibit stronger enhancements, which are not entirely due to the change of the nanoparticles' exposed areas. The absolute EF values calculated by using those bands intensities were of an order of 10^6. Such a band selective enhancement was previously observed for R6G and derives from the different kinds of molecules' adsorption sites (Kudelski 2005, Hildebrandt and Stockburger 1984). Thus, the existence of a dependence of the coupling between the localized and propagative modes on the silver nanoparticles' arrays on the nanoparticles vertical dimension (Baia et al. 2006c) was experimentally demonstrated.

The SERRS spectra of the adsorbed probe molecules (the R6G electronic absorption band is located at 520 nm) recorded with a 532 nm laser line are illustrated in Fig. 8.11*b*. One can see that they display the same features as those recorded off resonance, whereas a fluorescent background and differences in the relative intensities of some bands can be noticed. The existence of the fluorescence signal in the SERRS spectra is an initial hint of the enhancement due to the silver nanostructures' presence. The molecules situated directly on the substrate in the nanoparticles vicinity cannot contribute to the SERS signal and thus give rise to the fluorescent background. The obtained difference between the substrates' relative enhancement (1.4) and the nanoparticles exposed areas ratio (1.1) further confirms the existence of "hot edges" due to the combination of the particle plasmon excitation with long range photonic interactions occurring in arrays. The absolute EF values, found to be approximately 3 times higher than those obtained for off-resonance excitation, show the same nanoparticles' height dependent SERS enhancement trend, which proves the sensitivity of the electromagnetic couplings to the nanoparticles' out-of-plane dimension.

Thus, the experimental evidence of the electromagnetic couplings' sensitivity to the nanoparticles' vertical dimension was provided from the evaluation of the SERS efficiency of truncated tetrahedral silver nanoparticles with different heights

(Baia et al. 2006c) besides those found when the interparticles' distance (Gunnarsson et al. 2001), the simultaneously in-plane and out-of-plane particles' dimensions (Zou and Schatz 2005), and the concomitantly in-plane nanoparticles' dimension and the interparticles' distance (Schmidt et al. 2004) were varied. The unique contribution of the electromagnetic couplings between the LSPR of each particle and the photonic modes of the particle array to the SERS enhancement of the R6G molecules adsorbed on these substrates was demonstrated. Moreover, it was shown that the nanoparticles' height dependent SERS enhancement trend is maintained for the resonant excitation of the probe molecule.

In conclusion, we can emphasize that all of the above-mentioned SERS active substrates recommended by the large tunability of surface plasmon excitation, combined with the advantage of relatively high exhibited average EF values can be regarded as excellent candidates for forthcoming investigations of various pharmaceutically, biologically and medically relevant molecules.

References

Aizpurua J, Bryant GW, Richter LJ, García de Abajo FJ (2005) Optical properties of coupled metallic nanorods for field enhanced spectroscopy. Phys Rev B 71:235420-1–235420-13

Astilean S, Bolboaca M, Maniu D, Iliescu T (2003) Ordered metallic nanostructures for surface-enhanced Raman spectroscopy. Book of Abstracts of the 3rd Conference New Research Trends in Material Science, Constanta, 266

Astilean S, Bolboaca M, Maniu D, Iliescu T (2004a) Ordered metallic nanostructures for surface-enhanced Raman spectroscopy. Rom Rep Phys, 56:346–351

Astilean S, Baia M, Maniu D, Pinzaru S, Iliescu T (2004b) Fabrication of ordered noble-metal nanostructures via nanosphere lithography and their investigation as effective substrates for surface-enhanced Raman spectroscopy. Book of Abstracts of the International Bunsen Discussion Meeting "Raman and IR Spectroscopy in Biology and Medicine" Jena, 84

Astilean S, Baia M, Baia L, Farcau C, Toderas F (2005) Noble-metal films deposited on polystyrene colloidal crystal as effective substrate for surface-enhanced Raman spectroscopy. Book of Abstracts of the Surface Plasmon Photonics 2 Confererence, Graz, 124

Astilean S, Baia M, Maniu D, Baia L, Toderas F, Bica E, Iosin M, Popescu O, Craciun C, Barbu-Tudoran L, Socaciu C, Popp J, Baldeck PL (2006a) Bio-plasmonics: Sensing biomolecular interactions with gold nanoparticles. Book of Abstracts of the 2nd International Conference of Advanced Spectroscopies on Biomedical and Nanostructured Systems, Cluj-Napoca, 61

Astilean S, Baia M, Baia L, Farcau C, Maniu D (2006b) Tunable surface-enhanced Raman scattering (SERS) from noble metal films deposited on polystyrene colloidal crystal and nanoparticle arrays fabricated by nanosphere litography. Meeting Digest of the EOS Topical Meeting on Molecular Plasmonic Devices, Engelberg, 74–76

Atay T, Song JH, Nurmikko AV (2004) Strongly interacting plasmon nanoparticle pairs: from dipole-dipole interaction to conductively coupled regime. Nano Lett 4:1627–1631

Baia M, Toderas F, Mihut A, Baia L, Astilean S (2005a) Gold colloidal particles on functionalized glass substrate for optical sensing and SERS. Book of Abstracts of the 11th European Conference on Applications of Surface and Interface Analysis (ECASIA) Vienna, 161

Baia M, Baia L, Astilean S (2005b) Gold nanostructured films deposited on polystyrene colloidal crystal templates for surface-enhanced Raman spectroscopy. Chem Phys Lett 404:3–8

Baia M, Toderas F, Baia L, Popp J and Astilean S (2006a) Probing the enhancement mechanisms of SERS with p-aminothiophenol molecules adsorbed on self-assembled gold colloidal nanoparticles. Chem Phys Lett 422:127–132

Baia M, Toderas F, Baia L, Popp J, Astilean S (2006b) Monitoring the interplay of SERS enhancement mechanisms contribution with p-aminothiophenol adsorbed on self-assembled gold nanoparticles. Book of Abstracts of the 2nd International Conference of Advanced Spectroscopies on Biomedical and Nanostructured Systems, Cluj-Napoca, 122

Baia M, Baia L, Popp J, Astilean S (2006c) Surface-enhanced Raman scattering efficiency of truncated tetrahedral Ag nanoparticle arrays mediated by electromagnetic couplings. Appl Phys Lett 88:143121-1–143121-3

Baia M, Baia L, Popp J, Astilean S (2006d) Ordered metallic nanostructures obtained by nanosphere lithography as tunable SERS-active substrates. Book of Abstracts of the 2nd International Conference Advanced Spectroscopies on Biomedical and Nanostructured Systems, Cluj-Napoca, 123

Baia L, Baia M, Popp J, Astilean S (2006e) Gold films deposited over regular arrays of polystyrene nanospheres as highly effective SERS substrates from visible to NIR. J Phys Chem B 110:23982–23986

Billot L, Lamy de la Chapelle M, Grimault AS, Vial A, Barchiesi D, Bijeon JL, Adam PM, Royer P (2006) Surface enhanced Raman scattering on gold nanowire arrays: Evidence of strong multipolar surface plasmon resonance enhancement. Chem Phys Lett 422:303–307

Bolboaca M, Baia L, Chicinas I, Iliescu T, Astilean S (2003a) Optical and structural investigations of ordered metallic nanostructures for SERS experiments. Studia UBB Physica, XLVIII:357–359

Bolboaca M, Baia L, Chicinas I, Iliescu T, Astilean S (2003b) Optical and structural investigations of ordered metallic nanostructures for SERS experiments. Book of Abstracts of the 3rd Conference of Isotopic and Molecular Processes, Cluj-Napoca, 99

Campion A, Kambhampati P (1998) Surface enhanced Raman scattering. Chem Soc Rev 27:241–250

Felidj N, Aubard J, Levi G, Krenn JR, Salerno M, Schider G, Lamprecht B, Leitner A, Aussenegg FR (2002a) Controlling the optical response of regular arrays of gold particles for surface-enhanced Raman scattering. Phys Rev B 65:075419-1–075419-9

Felidj N, Aubard J, Levi G, Krenn JR, Schider G, Leitner A, Aussenegg FR (2002b) Enhanced substrate-induced coupling in two-dimensional gold nanoparticle arrays. Phys Rev B 66:245407-1–245407-7

Felidj N, Truong SL, Aubard J, Levi G, Krenn JR, Hohenau A, Leitner A, Aussenegg FR (2004) Gold particle interaction in regular arrays probed by surface enhanced Raman scattering. J Chem Phys 120:7141–7146

Grand J, Kostcheev S, Bijeon JL, Lamy de la Chapelle M, Adam PM, Rumyantseva A, Lerondel G, Royer P (2003) Optimization of SERS-active substrates for near-field Raman spectroscopy. Synth Met 139:621–624

Grand J, Lamy de la Chapelle M,. Bijeon JL, Adam PM, Vial A, Royer P (2005) Role of localized surface plasmons in surface-enhanced Raman scattering of shape-controlled metallic particles in regular arrays. Phys Rev B 72:033407-1–033407-4

Gunnarsson L, Bjerneld EJ, Xu H, Petronis S, Kasemo B, Käll M (2001) Interparticle coupling effects in nanofabricated substrates for surface-enhanced Raman scattering. Appl Phys Lett 78:802–804

Haynes CL, Van Duyne RP (2002) Plasmon scanned surface-enhanced Raman scattering excitation profiles. Mat Res Soc Symp Proc 728:S10.7.1–S10.7.6

Haynes CL, van Duyne RP (2003) Plasmon-Sampled Surface-Enhanced Raman Excitation Spectroscopy. J Phys Chem B 107:7426–7433

Haynes CL, McFarland AD, Van Duyne RP (2005) Surface enhanced Raman spectroscopy. Anal Chem 340:339–346

Hildebrandt P, Stockburger M (1984) Surface-enhanced resonance Raman spectroscopy of Rhodamine 6G adsorbed on colloidal silver. J Phys Chem 88:5935–5944

Hu X, Wang T, Wang L, Dong S (2007) Surface-enhanced Raman scattering of 4-aminothiophenol self-assembled monolayers in sandwich structure with nanoparticle shape dependence: Off-surface plasmon resonance condition. J Phys Chem C 111:6962–6969

Hulteen JC, Van Duyne RP (1995) Nanosphere lithography: a materials general fabrication process for periodic particle array surfaces. J Vac Sci Technol A 13:1553–1558

Jensen TR, Malinsky MD, Haynes CL, Van Duyne RP (2000) Nanosphere lithography: Tunable localized surface plasmon resonance spectra of silver nanoparticles. J Phys Chem B 104:10549–10556

Kudelski A (2005) Raman studies of rhodamine 6G and crystal violet sub-monolayers on electrochemically roughened silver substrates: Do dye molecules adsorb preferentially on highly SERS-active sites? Chem Phys Lett 414:271–275

Li X, Xu W, Zhang J, Jia H, Yang B, Zhao B, Li B, Ozaki Y (2004) Self-assembled metal colloid films: Two approaches for preparing new SERS active substrates. Langmuir 20:1298–1304

Li Y, Zhou J, Zhang K, Sunb C (2007) Gold nanoparticle multilayer films based on surfactant films as a template: Preparation, characterization, and application. J Chem Phys 126:094706-1–094706-7

McFarland AD, Young MA, Dieringer JA, van Duyne RP (2005) Wavelength-scanned surface-enhanced Raman excitation spectroscopy. J Phys Chem B 109:11279–11285

Moskovits M (2005) Surface-enhanced Raman spectroscopy: a brief retrospective. J Raman Spectrosc 36:485–496

Orendorff CJ, Gole A, Sau TK, Murphy CJ (2005) Surface-enhanced Raman spectroscopy of self-assembled monolayers: Sandwich architecture and nanoparticle shape dependence. Anal Chem 77:3261–3266

Schmidt JP, Cross SE, Buratto SK (2004) Surface-enhanced Raman scattering from ordered Ag nanocluster arrays. J Chem Phys 121:10657–10659

Tamaru H, Kuwata H, Miyazaki HT, Miyano K (2002) Resonant light scattering from individual silver nano-particles and particle pairs. Appl Phys Let 80:1826–1828

Tian ZQ, Ren B, Wu DY (2002) Surface-enhanced Raman scattering: From noble to transition metals and from rough surfaces to ordered nanostructures. J Phys Chem B 106:9463–9483

Toderas F, Mihut A M, Baia M, Simon S, Astilean S (2004a) Self-assembled gold nanoparticles on solid substrate. Studia UBB Physica XLIX:89–94

Toderas F, Mihut A, Baia M, Astilean S (2004b) Self-assembling noble metal nanoparticles and their investigation for applications in surface-enhanced Raman spectroscopy (SERS) and surface plasmon resonance (SPR) biodetection. Book of Abstracts of the 1st International Conference on Advanced Spectroscopies on Biomedical and Nanostructured Systems, Cluj-Napoca, 96

Toderas F, Boca S, Baia M, Baia L, Maniu D, Astilean S, Simon S (2006a) Self-assembled multilayers of gold nanoparticles as versatile platforms for molecular sensing by Fourier transform-surface enhanced scattering (FT-SERS) and surface enhanced infrared absorption (SEIRA). Book of Abstracts of the 2nd International Conference Advanced Spectroscopies on Biomedical and Nanostructured Systems, Cluj-Napoca, 121

Toderas F, Baia M, Baia L, Maniu D, Farcau C, Astilean S, Barbu-Tudoran L, Craciun C (2006b) Gold nanoparticles self-assembled on functionalized glass substrates and their surface plasmons enhanced properties. Book of Abstracts of the International Conference of Micro to Nanophotonics, ROMOPTO, Sibiu, 45

Toderas F, Baia M, Baia L, Astilean S (2007) Controlling gold nanoparticle assemblies for efficient surface enhanced Raman scattering (SERS) and localized surface plasmon resonance (LSPR) sensors. Nanotechnology, 18:doi:10.1088/0957–4484/18/25/255702

Zhang X, Hicks EM, Zhao J, Schatz GC, van Duyne RP (2005) Electrochemical tuning of silver nanoparticles fabricated by nanosphere lithography. Nano Letters 5:1503–1507

Zhou Q, Zhao G, Chao Y, Li Y, Wu Y, Zheng J (2007) Charge-transfer induced surface-enhanced Raman scattering in silver nanoparticle assemblies. J Phys Chem C 111:1951–1954

Zhou Q, Li X, Fan Q, Zhang X, Zheng J (2006) Charge transfer between metal nanoparticles interconnected with a functionalized molecule probed by surface-enhanced Raman spectroscopy. Angew Chem Int Ed 45:3970–3973

Zou S, Schatz GC (2005) Silver nanoparticle array structures that produce giant enhancements in electromagnetic fields. Chem Phys Lett 403:62–67

Wang W, Gu B (2005) New surface-enhanced Raman spectroscopy substrates via self-assembly of silver nanoparticles for perchlorate detection in water. Appl Spectrosc 59:1509–1515

Wei H, Li J, Wang Y, Wang E (2007) Silver nanoparticles coated with adenine: preparation, self-assembly and application in surface-enhanced Raman scattering. Nanotechnology 18:175610-1–175610-5

9 Summary

Throughout this book, the results of our recent infrared, Raman, and SERS spectroscopic investigations of some molecules of pharmaceutical interest have been presented. We described the spectroscopic analysis of some tranquilizers, sedatives, anti-inflammatory drugs, drugs with antibacterial properties, vitamins, and other molecules with pharmacological activity.

In order to understand the adsorption behavior of the molecules on the colloidal metallic particles, detailed vibrational investigations using infrared and Raman spectroscopy in conjunction with theoretical calculations were performed. The results of theoretical calculations established the most stable conformation of the molecular species and provided the theoretical wavenumber values employed for the assignment of the experimentally determined vibrational modes.

Having in mind these considerations, the fundamentals of infrared and Raman spectroscopy, SERS, and theoretical simulations were briefly described in the second chapter. The chapter starts by presenting details about the molecular vibrations and further illustrates the basics of infrared and Raman spectroscopy. The SERS effect is further explained and the mechanisms of the surface enhancement are briefly described. The chapter ends with a survey of the SERS-active substrates developed over time. A short description of the methods employed in theoretical simulations of the structure and vibrational wavenumbers of the investigated species is also included in the last subchapter. Information about the experimental measurements and computational simulations are also specified in this chapter.

The SERS spectra analyses of the molecules investigated in this book provided information concerning the structure of the adsorbed species and their orientation relative to the metal surface. Furthermore, the pH influence on the adsorption behavior of the investigated molecules has been monitored.

In the third chapter the SERS investigations performed on two types of drugs, e. g., tranquilizers (phenothiazine derivatives) and sedatives (anthranil) are illustrated. In the case of 10-isopentyl-10H-phenothiazine-5-oxide and 10-isopentyl-10H-phenothiazine-5,5-dioxide derivatives prior to analyzing the SERS spectra, the conformation of the most stable isomers was theoretically predicted and their vibrational investigation was further carried out. The SERS spectra revealed that both molecules are chemisorbed on the metal surface through the oxygen atom, and the changes observed in the SERS spectra at different pH values are due to a reorientation of the adsorbed molecule with respect to the silver surface. In the case of anthranil, by correlating the spectroscopic changes evidenced between the Raman and SERS spectra and the results of DFT calculations performed on different Ag-anthranil model complexes, it was concluded that the anthranil molecule is adsorbed on the colloidal silver surface through the lone pair electrons of the nitrogen atom and has a titled orientation relative to the metal surface.

The next chapter shows the study of the interaction between the diclofenac sodium (DCFNa), a well-known anti-inflammatory agent, and β-cyclodextrin (βCD) in the solid state guest-host inclusion complex by using Raman and SERS. Firstly, a detailed vibrational analysis of the most stable conformer of the free DCFNa, as predicted from theoretical simulations, was performed. The SERS spectra obtained in acidic and neutral environments indicated the chemisorption of the DCF molecule on the silver surface via the lone pair oxygen electrons of the carboxylate group, which has a perpendicular or slightly tilted orientation with respect to the silver surface. By analyzing SERS spectra at different pH values, a change of the phenyl rings' orientation with respect to the metal surface from a tilted close to flat to a more perpendicular one was also concluded. Furthermore, from the analysis of the Raman spectra of the guest-host encapsulation complex interactions between both the dichlorophenyl ring and the phenylacetate group of the DCFNa species and βCD molecule were evidenced. The changes in the SERS spectra recorded at different pH values demonstrated that, depending on the pH values, different isomeric forms of the guest-host complex are preferentially adsorbed on the silver surface, but the adsorption of the guest molecule is maintained in both cases through the nonbonding electrons of the oxygen atom. The probable orientation of the adsorbed species relative to the silver surface was also indicated.

Throughout the fifth chapter the adsorption behavior of a few molecules with antibacterial properties is elucidated by using the SERS technique. Thus, the first two subchapters describe the investigations of potassium benzylpenicillin (KBP) and trihydrate amoxicillin (THA), whereas in the last subchapter the rivanol molecule is examined. The analysis of the SERS spectrum of the KBP recorded on silver colloid near neutral environment revealed the chemisorption on the metal surface via the carboxylate anion. The adsorption of the KBP molecule on the silver surface takes place in such a way that the phenyl, beta-lactam, and thiazolidine rings are located at a relatively large distance to the silver surface and are oriented approximately perpendicularly to it. Moreover, the SERS spectrum of THA recorded on silver colloidal suspension at the pH value of 6 evidenced the chemisorption of the THA molecules, which are bonded on the silver surface through the amino group. It was also found that the phenyl, beta-lactam, and thiazolidine rings of the THA species are located at a relatively large distance with respect to the silver surface. On the other hand, the SERS spectra of rivanol revealed the physisorption of monocation species on the silver surface.

The following chapter describes the study of the adsorption behavior of B1 and PP vitamin at different pH values carried out by SERS. Prior to evaluating the adsorption behavior, the pH-dependence Raman spectra of the thiamine (B1 vitamin) aqueous solution was accomplished. The presence of two molecular species, the protonated and neutral form, was revealed and the pK_a value for the protonation of a nitrogen atom of the pyrimidine ring was found to be slightly over 5. In a strong alkaline environment (pH > 8) the denaturation of the molecule was observed. Moreover, the pH-dependence SERS study of B1 vitamin revealed the presence of two different adsorbed molecular species and their coexistence. A higher adsorption affinity to the gold surface of the neutral molecular form was concluded, as the adsorption of the neutral thiamine species was observed at pH values lower with two units than the pK_a value of the molecule. The chemisorption of the thiamine molecules takes place through the nitrogen atoms of the pyrimidine ring, the orientation of the molecule with respect to the gold surface depending on the protonation degree. On the other hand, it was established that the SERS spectra of PP vitamin are preponderantly due to the neutral form at pH values equal with and above 5.5, and to the protonated form at acidic pH. It was also found that the protonated form is adsorbed via an amide group, while the neutral form is bound through the ring nitrogen atom.

Throughout the seventh chapter the attention was focused on the investigation of other molecules with pharmacological activity, including 2-formylfuran (2FF), quinoline, and acridine derivatives. In the first subchapter the investigations of 5-(4-fluor-phenyl)-2-formylfuran and 5-(4-brom-phenyl)-2-formylfuran derivatives together with those of the 2FF are presented. Prior to analyzing their adsorption behavior, a detailed vibrational investigation of the most stable conformer was carried out. The SERS spectrum of 2FF shows that the molecules are chemisorbed on the silver surface through both the ring oxygen and the oxygen atom of the substituent group, the *cis*-form being preferred in the adsorbed state. The adsorbed molecules are oriented perpendicularly, or at least tilted, with respect to the silver surface. From the SERS spectra of 2FF derivatives it was concluded that their chemisorption is via the nonbonding electrons of the ring oxygen and they have a perpendicular orientation, or at least one tilted with respect to the silver surface. In the case of quinoline derivatives, it was found that the SERS spectra are strongly dependent on the pH conditions. Thus, the isoquinoline molecules are adsorbed via the nonbonding electrons of the nitrogen as neutral molecules, the variation in SERS spectra at different pH values being attributed to a change in orientation of adsorbed molecules. In contrast, the lepidine molecules at an acidic pH are preferentially adsorbed on the silver surface in the protonated form, via the π^* electrons of the ring. At an alkaline pH the adsorbed molecules are tilted orientated to the metal surface and bonded to it via the lone pair of the nitrogen. Finally, the analysis of the SERS spectra of neutral 9-phenylacridine reveals its adsorption on the silver surface via the lone pair electrons of the nitrogen atom. At acidic pH values the 9-phenylacridinium cation pairs with the chloride ion and this ion pair is bound to the silver surface via the chloride ion.

Bearing in mind that each drug is specific to a certain human organ on which it is adsorbed on some special centers, as revealed from pharmacological studies, the adsorption of the molecules on a metal surface can be regarded as a mimic of this adsorption process. Moreover, to understand the action of potential drugs, it is very important to know if the structure of the adsorbed species is the same as that of the free molecule, and also it is important to establish whether or not the molecule-substrate interaction may be dependent on the pH value of the environmental solution. Thus, after elucidating the adsorption mechanism of a molecule on a silver or gold surface that may serve as an analogue for an artificial biological interface, the study can be expanded to the adsorption on membranes or other interesting biological surfaces for medical or therapeutic treatments.

In addition, since there is a huge interest concerning how to control, manipulate, and amplify light on the nanometer length scale using the surface plasmons properties, a few examples related to the developments in SERS-active substrates are given in the eighth chapter. Thus, a few substrates consisting of self-assembled colloidal nanoparticles on functionalized substrates, corrugated gold films deposited on top of highly ordered polystyrene nanospheres arrays, and truncated tetrahedral silver nanoparticles arrays were analyzed from the perspective of their SERS efficiency by employing test molecules. The developed SERS-active substrates recommended by the large tunability of surface plasmon excitation, combined with the advantage of relatively high exhibited average enhancement factors, can be regarded as excellent candidates for forthcoming investigations of various pharmaceutically, biologically, and medically relevant molecules.

Index

Printing: Krips bv, Meppel, The Netherlands
Binding: Stürtz, Würzburg, Germany